Electronically Scanned Arrays

MATLAB® Modeling and Simulation

Edited by **Arik D. Brown**

CRC Press
Taylor & Francis Group
Boca Raton   London   New York

CRC Press is an imprint of the
Taylor & Francis Group, an **informa** business

CRC Press
Taylor & Francis Group
6000 Broken Sound Parkway NW, Suite 300
Boca Raton, FL 33487-2742

First issued in paperback 2017

© 2012 by Taylor & Francis Group, LLC
CRC Press is an imprint of Taylor & Francis Group, an Informa business

No claim to original U.S. Government works

Version Date: 20120309

ISBN 13: 978-1-4398-6163-9 (hbk)
ISBN 13: 978-1-138-07403-3 (pbk)

Visit the Taylor & Francis Web site at
http://www.taylorandfrancis.com

and the CRC Press Web site at
http://www.crcpress.com

Dedication

I dedicate this to the Lord, who gave me the inspiration to write this book; to my wonderful wife, Nadine, who has never faltered in her belief in my abilities; and to my two children, Alexis and Joshua, whom I hope this will inspire to fulfill their purpose and destiny in life.

Contents

Preface

During the course of my career at Northrop Grumman, I have had the pleasure of teaching several internal classes in addition to mentoring young engineers new to the company. Several years ago one of the things I noticed was that many of our talented new employees were lacking the insight that comes from understanding the fundamentals. I often saw antenna analysis performed with no real insight into key concepts for understanding and students tackling more difficult problems with no textbook solution. It is widely known that an electronically scanned array (ESA) pattern can be calculated by simply typing in the *fft* function in MATLAB®, and, in the blink of an eye you have before you a beautifully colored ESA pattern. However, when it comes to analyzing subarray architectures, conformal arrays, etc., it is imperative that the fundamentals of an ESA are known— as opposed to how to type the word *fft* or *fftshift* in MATLAB. One of the goals of this book is to provide a vehicle whereby those who are new to ESAs can get a fundamental understanding of how they work, and also to provide code to run as an aid without having to start from scratch.

Several years ago I had the honor and privilege of taking the reins of an ESA course that was being given to customers who were unfamiliar with ESA technology, and who might not understand some of the fundamental concepts. One of my mentors, Bill Hopwood, was the originator of this course, and it has proven quite beneficial in educating new customers on our ESA expertise at Northrop Grumman. I also wanted this book to be a reference that I could refer to those who had an interest in learning more about ESAs and wanted more than just a cursory top-level introduction.

Last, this book is intended to serve as both a textbook in a college/university setting as well as a reference for practicing engineers. MATLAB is one of the best tools available for exploring and understanding science and engineering concepts. The code provided within this book is a great vehicle for understanding the analysis and simulation of ESAs.

This book is comprised of six chapters. Chapters 1 and 2 deal primarily with the fundamentals of ESA theory that should be a part of the knowledge base of anyone who is serious about ESAs. Concepts such as beamwidth, grating lobes, instantaneous bandwidth, etc., are covered in detail. Chapter 1 focuses solely on one-dimensional ESAs, while Chapter 2 delves into two-dimensional ESA analysis and includes coordinate definitions, sine space, and pattern analysis.

Chapters 3 and 4 build upon the fundamental ESA topics covered in Chapters 1 and 2, and advances to subarrays and pattern optimization. Chapter 3 focuses on subarray architectures and the nuances involved with subarrayed architectures, such as phase vs. time delay steering, instantaneous bandwidth considerations, and the utility of digital beamforming. The chapter concludes with the analysis of overlapped subarrays.

Chapter 4 was written by my colleague, Dr. Daniel Boeringer. It focuses on pattern optimization. Developing methodologies for optimizing the spatial distribution of sidelobes is of great importance, and this chapter provides an introduction to pattern optimization and the added capability it provides.

Chapters 5 and 6 cover topics that have not been explored as thoroughly, if at all, by other books in regard to ESAs. Presenting this information is extremely exciting and I think it will be of great benefit to the reader. Chapter 5 focuses on the application of ESAs in a space environment. This chapter was written by an extremely talented system engineer, Timothy Cooke. The chapter focuses on pattern projections, translation of sine space to latitude/longitude coordinates for ESAs operating in an orbital regime, and covers some basic astrodynamics principles that are necessary to understand when analyzing space-based applications for ESAs.

Chapter 6 concludes with reliability aspects, which are quite frankly, often overlooked when it comes to analyzing ESAs. The material in this chapter is based primarily on a technical memo authored by one of Northrop's retired senior system architects, Bill Hopwood. My colleague Dr. Jabberia Miller formalized the notes and has provided a most useful chapter on reliability analysis and its impact on ESA performance.

MATLAB® is a registered trademark of The MathWorks, Inc. For product information, please contact:

The MathWorks, Inc.
3 Apple Hill Drive
Natick, MA 01760-2098 USA
Tel: 508 647 7000
Fax: 508-647-7001
E-mail: info@mathworks.com
Web: www.mathworks.com

Acknowledgments

First, I thank God for giving me the ability, opportunity, and the inspiration to create this book. Late on a Tuesday night, while driving from class at Spirit of Faith Bible Institute in Temple Hills, Maryland, the Lord gave me the idea to write this book. I had no intentions or previous aspirations to author a book, and it has truly been a God-inspired journey.

I thank my wife Nadine for her prayers, support, and sacrifice of time while I completed this book. I would not be the man I am today without her. I also want to thank my daughter and son, Alexis and Joshua, for sacrificing "daddy time" when I had to work on the book. To my parents, Rudy and Meredith, thank you for always believing in, and supporting me. Your guidence, since my childhood, paved the way to my education at MIT and the University of Michigan, to my success at Northrop Grumman. The sacrifices you made for me can never repaid. To my grandparents, James and Ethel Mae, Wesley and Lilly, my godmother Ruth, my aunt Deborah, my mother-in-law and father-in-law Raymond and Kathleen Stuart, I thank you for providing the extra support and encouragement. To my extended family (and there are many), I thank you as well.

I must thank three outstanding gentlemen who each wrote a chapter in this book, Dr. Daniel Boeringer, Dr. Jabberia Miller, and Timothy Cooke. It has been a pleasure being your colleague. You are truly some of Northrop Grumman's best and brightest. I also thank all the individuals who have helped to make this book a reality: John Wojtowicz, Richard Konapelsky, Kurt Ramsey, Kevin Idstein, Urz Batzel, Dr. Sumati Rajan, Jon Ulrey, John Long, Kevin Schmidt, and Tom Harmon. Many thanks to Dr. Joseph Payne for taking the time to review this manuscript, and to my past advisor, Dr. John L. Volakis, who has always had an open door if I needed to ask him questions. Thanks for being an outstanding example of a great work ethic. Last, thanks to Dr. Leo Kempel.

There are three individuals who have been both friends and great mentors in my career Northrop Grumman. Mentors lead by example and allow the mentee to follow and learn along the way. From these three people I learned that intellect without experience and hard work really doesn't amount to much. To Bill Hopwood, affectionately known as

"Uncle Bill" and "Mr. ESA," I thank you for being an advocate for me. You are truly one of the most innovative people I have ever met, and it has been a pleasure working with you. To Lee Cole, thank you for showing me the ropes in regard to being a system architect, and for your open-door policy. You are truly an amazing individual from whom I am still learning. To Dan Davis, I thank you for all the one-on-one tutorials, and your willingness to always talk about concepts whenever I visited your office. I learned from you that many times, difficult problems can be simplified with short, elegant solutions. Early in my career you instructed me to keep the MATLAB® codes I wrote in my own "toolbox" to have them ready to use. Well, I followed your instructions!

Finally, I would be remiss if I didn't thank my pastor and spiritual father, Dr. Michael A. Freeman. I have learned from you to lean on God first and then lean on my intellect second (which is not always easy!). Thank you for your leadership, exemplary lifestyle, and pushing me to never settle for being average.

Contributors

Daniel Boeringer, PhD
Northrop Grumman Electronic
  Systems
Baltimore, Maryland

Arik D. Brown, PhD
Northrop Grumman Electronic
  Systems
Baltimore, Maryland

Timothy Cooke
Northrop Grumman Electronic
  Systems
Baltimore, Maryland

Jabberia Miller, PhD
Northrop Grumman Electronic
  Systems
Baltimore, Maryland

chapter one

Electronically Scanned Array Fundamentals—Part 1

Arik D. Brown
Northrop Grumman Electronic Systems

Contents

1.1 Introduction

Electronically scanned arrays (ESAs) provide the capability of command-able, agile, high-gain beams, which is advantageous for applications such as radar, weather surveillance, and imaging. In contrast to reflector anten-nas, which require a gimbal for steering the array beam, ESAs electroni-cally scan the array beam in space without physical movement of the array (see Figure 1.1). Scanning the beam with an ESA can be performed on the order of microseconds as opposed to milliseconds for a reflector. This enables increased scanning rates in addition to flexibility to command the array beam's position in space for tailored modes of operation.

When designing ESAs, there are basic fundamentals that need to be understood for a successful design (i.e., grating lobes, beamwidth, instan-taneous bandwidth, etc.). Additionally, a fundamental understanding of the application of ESAs is necessary (i.e., pattern optimization, subarrays, digital beamforming (DBF), etc.). After these foundations are set, then an understanding of practical aspects of ESAs, such as reliability, should

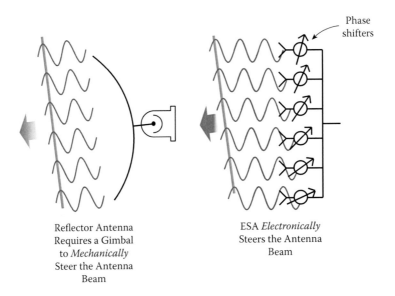

Reflector Antenna
Requires a Gimbal
to *Mechanically*
Steer the Antenna
Beam

ESA *Electronically*
Steers the Antenna
Beam

Phase
shifters

Figure 1.1 Reflector vs. ESA steering. (From Walsh, T., et al., *Active Electronically Scanned Antennas (AESA)*, Radar Systems Engineering Course, Northrop Grumman Electronic Systems, Baltimore, MD, December 2003.)

provide a well-rounded understanding of designing ESAs. The remainder of this chapter will focus on basic ESA fundamentals.

1.2   General One-Dimensional Formulation

1.2.1   Pattern Expression without Electronic Scanning

Consider a one-dimensional array of M elements as shown in Figure 1.2. The elements are uniformly spaced with a spacing of d. The overall length of the array, L, is equal to Md. The elements are centered about $x = 0$, and their position can be denoted as

$$x_m = (m - 0.5(M + 1))d, \text{ where } m = 1, \ldots, M \tag{1.1}$$

Each element has a complex voltage denoted as a_m. A signal that is incident on the array from a direction θ is captured by each of the array elements and is then summed together coherently for the composite signal. The expression for the coherent sum of voltages is represented as

$$AF = \sum_{m=1}^{M} A_m e^{j \frac{2\pi}{\lambda} x_m \sin \theta} \tag{1.2}$$

AF is the array factor and describes the spatial response of the M array elements. From Equation (1.2), we can see that the AF is a function of the aperture distribution (A_m), frequency ($\lambda = \frac{c}{f}$, c is the speed of light and f is frequency), element spacing (d), and angle (θ). The AF in Equation (1.2) has a maximum value when $\theta = 0°$. This maximum value is M, which is the

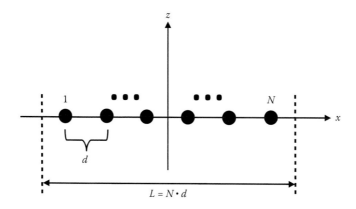

Figure 1.2 Linear array of N elements.

number of elements in the array. When writing code in MATLAB® this is a good check to ensure that no errors are in the code. As we'll show later, regardless of whether the array is one-dimensional or two-dimensional, the maximum value of the AF is always equal to the number of elements in the array.

The AF does not completely describe the spatial response of the array. Each of the elements in the array has an element pattern that is the elements spatial response. A good expression for modeling the element pattern is a cosine function raised to a power that is called the element factor (*EF*). The expression for the element pattern (*EP*) is

$$EP = \cos^{\frac{EF}{2}} \theta \tag{1.3}$$

In real applications, the *EP* does not go to zero at $\theta = 90°$. An ESA, in its installed environment or in a measurement range, will be subject to diffraction and reflections near the edges of the array that will modify the *EP* near the edges.

The complete pattern representation for the array is found using pattern multiplication. Pattern multiplication states that the complete pattern can be calculated by multiplying the AF and EP. This assumes that the EP is identical for each element in the ESA, which for large ESAs is a good characterization. Equation (1.4) shows the total pattern for the array of *M* elements.

$$F(\theta) = EP \cdot AF = \cos^{\frac{EF}{2}} \theta \cdot \sum_{m=1}^{M} A_m e^{j\frac{2\pi}{\lambda} x_m \sin \theta} \tag{1.4}$$

Several important points can be gleaned from Equation (1.4). The first is that the expression for the *EP* has been factored out of the AF under the assumption that the element pattern is the same for all elements. As we will see later in the chapter, this is not true for conformal arrays where the normal of the element pattern is not parallel with the normal of the array boresite ($\theta = 0°$), and is also incorrect for arrays with a small number of elements. However, outside of the discussion on conformal arrays, all ESAs described will assume a large number of elements. Another point that can be overlooked is that *EF* is divided by 2 in the expression for *EP*. This is because the element pattern power is EP^2 and *EP* is the voltage representation which is the square root of the power expression.

1.2.2 *Pattern Expression with Electronic Scanning*

In the previous section, a general result was shown for the spatial pattern representation of a linear *M* element array. In this section we will express the pattern with the inclusion of electronic scan. The expression in Equation (1.4) only has a maximum value when $\theta = 0°$. An ESA has the ability to scan the beam so that the beam has a maximum at other angles ($\theta \neq 0°$). For the remainder of this book the scan angle will be denoted as θ_o.

Scanning the beam of the array requires adjusting the phase or time delay of each element in the array.* By rewriting the expression in Equation (1.2) and expanding the complex voltage at each element ($A_m = a_m e^{j\Theta m}$) Equation (1.2) becomes

$$AF = \sum_{m=1}^{M} a_m e^{j\Theta_m} e^{j\frac{2\pi}{\lambda} x_m \sin\theta}$$

(1.5)

The AF then has a maximum at θ_o when $\Theta_m = -\frac{2\pi}{\lambda} x_m \sin\theta_o$ Equation (1.5) can then be expressed as

$$AF = \sum_{m=1}^{M} a_m e^{j\left(\frac{2\pi}{\lambda} x_m \sin\theta - \frac{2\pi}{\lambda} x_m \sin\theta_o\right)}$$

(1.6)

By applying the appropriate phase at each element, the ESA beam can be moved spatially without physically moving the entire array. This is the excitement of ESAs! The overall pattern can now be expressed as

$$F(\theta) = \cos^{\frac{EF}{2}}\theta \cdot \sum_{m=1}^{M} a_m e^{j\left(\frac{2\pi}{\lambda} x_m \sin\theta - \frac{2\pi}{\lambda} x_m \sin\theta_o\right)}$$

(1.7)

Electronic scan can be categorized as phase steering or time delay steering. For phase steering, each element has a phase shifter and applies the appropriate phase as a function of frequency and scan angle. A characteristic of phase shifters is that their phase delay is designed to be constant over frequency. This means the expression in Equation (1.7) must be modified to account for this. The pattern expression for phase delay steering becomes

$$F(\theta) = \cos^{\frac{EF}{2}}\theta \cdot \sum_{m=1}^{M} a_m e^{j\left(\frac{2\pi}{\lambda} x_m \sin\theta - \frac{2\pi}{\lambda_o} x_m \sin\theta_o\right)}$$

(1.8)

---

* For ESAs that employ both phase and time delay, both forms of delay must be adjusted. This applies to subarrayed ESAs, which will be discussed in Chapter 3.

where $\lambda = \frac{c}{f}$ and $\lambda_o = \frac{v}{f_o}$. What is readily seen is that when $f \neq f_o$ the pattern is no longer a maximum. This will be discussed in more depth in Section 1.3.2. When time delay is used Equation (1.7) becomes

$$F(\theta) = \cos^{\frac{EF}{2}} \theta \cdot \sum_{m=1}^{M} a_m e^{j\frac{2\pi}{\lambda} x_m (\sin\theta - \sin\theta_o)} \tag{1.9}$$

1.3   ESA Fundamental Topics

The AF for a one-dimensional ESA was shown to be

$$AF = \sum_{m=1}^{M} a_m e^{j\left(\frac{2\pi}{\lambda} x_m \sin\theta - \frac{2\pi}{\lambda} x_m \sin\theta_o\right)} \tag{1.10}$$

Mathematically, Equation (1.10) can be represented in a closed-form solution. The alternate expression for the AF is

$$AF = \frac{\sin\left[M\pi d\left(\frac{\sin\theta_o}{\lambda_o} - \frac{\sin\theta}{\lambda}\right)\right]}{\sin\left[\pi d\left(\frac{\sin\theta_o}{\lambda_o} - \frac{\sin\theta}{\lambda}\right)\right]} \tag{1.11}$$

Equation (1.11) is the phase shifter representation of the AF. The corresponding time delay formulation is

$$AF = \frac{\sin\left[\frac{M\pi d}{\lambda}(\sin\theta_o - \sin\theta)\right]}{\sin\left[\frac{\pi d}{\lambda}(\sin\theta_o - \sin\theta)\right]} \tag{1.12}$$

The derivation for the closed-form solution of the AF can be found in Appendix 1. The next three topics covered are readily derivable from Equations (1.11) and (1.12).

1.3.1   Beamwidth

The ESA beamwidth refers to the angular extent of the main beam where the power drops by a certain value. When that value is 3 dB, then the beamwidth can also be referred to as the half-power beamwidth. The common expression used for back of the envelope calculations we will show is actually the 4 dB beamwidth for a uniform distribution.

Equation (1.12) has the form of $\frac{\sin Mx}{\sin x}$, which can be approximated by $\frac{\sin x}{x}$, the standard sinc function. The 4 dB point for the sinc function occurs when $x = \pm \frac{\pi}{2}$. When $x = \pm \frac{\pi}{2}$, $\frac{\sin x}{x} = \frac{2}{\pi}$. With this same logic the expression in Equation (1.12) can be written as

$$AF \approx \frac{\sin\left[M\pi d\left(\frac{\sin\theta_o}{\lambda_o} - \frac{\sin\theta}{\lambda}\right)\right]}{M\pi d\left(\frac{\sin\theta_o}{\lambda_o} - \frac{\sin\theta}{\lambda}\right)} \tag{1.13}$$

Setting the arguments in Equation (1.13) equal to $\pm\frac{\pi}{2}$ when $\theta = \theta_o \pm \frac{\theta_{BW}}{2}$ produces the following two equations:

$$M\pi d\left(\frac{\sin\theta_o}{\lambda_o} - \frac{\sin\left(\theta_o + \frac{\theta_{BW}}{2}\right)}{\lambda}\right) = \frac{\pi}{2} \tag{1.14}$$

and

$$M\pi d\left(\frac{\sin\theta_o}{\lambda_o} - \frac{\sin\left(\theta_o - \frac{\theta_{BW}}{2}\right)}{\lambda}\right) = -\frac{\pi}{2} \tag{1.15}$$

Subtracting Equation (1.15) from Equation (1.14) and applying a trigonometric identity produces the expression

$$\frac{M\pi d}{\lambda}\left(2\sin\frac{\theta_{BW}}{2}\cos\theta_o\right) = \pi \tag{1.16}$$

Using the sine small angle approximation, Equation (1.16) can be solved for the beamwidth and is expressed as

$$\theta_{BW} = \frac{\lambda}{Md\cos\theta_o} = \frac{\lambda}{L\cos\theta_o} \tag{1.17}$$

where $L = Md$, and is the length of the ESA aperture. Equation (1.17) is valid for both phase shifter and time delay steering, and for $\theta_o = 0$ is the typical equation used to estimate ESA beamwidth. From Equation (1.17) we can see that the beamwidth is inversely proportional to frequency, aperture length, and the cosine of the scan angle. The beamwidth in Equation (1.17)

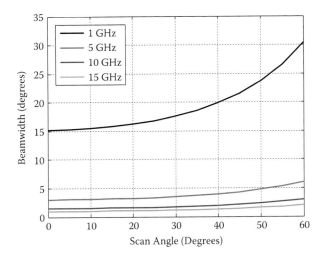

Figure 1.3 ESA beamwidth as a function of scan angle and frequency for $k = 1$.

is the 4 dB beamwidth for a uniform aperture illumination. A more general expression for the beamwidth is

$$\theta_{BW} = \frac{k\lambda}{L\cos\theta_o} \qquad (1.18)$$

where k is the beamwidth factor and varies depending on the aperture distribution. As an example, $k = 0.886$ for the 3 dB beamwidth of a uniformly illuminated ESA. Figure 1.3 shows a plot of ESA beamwidth as a function of scan angle and frequency for $k = 1$.

1.3.2 Instantaneous Bandwidth

When describing instantaneous bandwidth (IBW) it is advantageous to first think of it from a phase shifter perspective. For an ESA employing phase delay, the phase shifters are set at each element to scan the beam. The phase shifters have the characteristic of constant phase vs. frequency. At the tune frequency of the ESA, $f = f_o$ in Equation (1.11) and the AF has a maximum value. When $f = f_o + \Delta f$, the AF no longer has a maximum value at $f = f_o$ and there is an associated pattern loss at the commanded scan angle. This phenomenon is commonly referred to as beam squint. The IBW is the range of frequencies over which the loss is

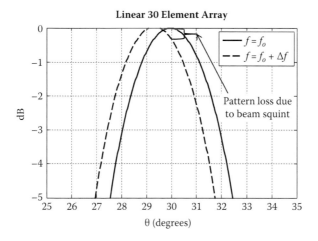

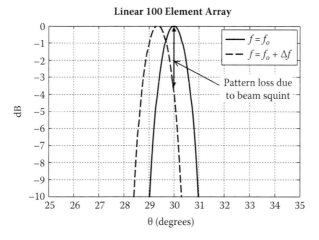

Figure 1.4 Beam squint using phase-shifter steering for ESAs with the same element spacing with differing number of elements. In both plots the solid line pattern represents operation at the tune frequency f_o and the dashed line represents beam squint at $f = f_o + \Delta f$.

acceptable and is $2\Delta f$. Typically, the IBW specified is the 3 or 4 dB IBW. Figure 1.4 illustrates beam squint for two ESAs with different aperture lengths. From the plots in Figure 1.4 we can see that the array with the longer length has a smaller beamwidth and suffers more loss at the commanded scan angle of 30°.

The IBW can be derived from Equation (1.13) in a similar fashion to the array beamwidth. (An alternate derivation using the exponential form of the AF can be found in Appendix 2.) The first step is to express Equation (1.13) in terms of frequency:

$$AF \approx \frac{\sin\left[\frac{M\pi d}{c}(f_o \sin\theta_o - f\sin\theta)\right]}{\frac{M\pi d}{c}(f_o \sin\theta_o - f\sin\theta)}$$

(1.19)

Substituting $f = f_o + \Delta f$ into Equation (1.19) and solving for Δf, the following expression is obtained:

$$IBW = \Delta f = \frac{c}{Md\sin\theta_o} = \frac{c}{L\sin\theta_o}$$

(1.20)

Equation (1.20) defines the 4 dB IBW for an ESA. Similar to Equation (1.18) the IBW can also be written more generally as

$$IBW = \frac{kc}{L\sin\theta_o}$$

(1.21)

where k is the beamwidth factor, which is a function of the aperture distribution.

For the case of time-delay steering there is no beam squint. Looking at Equation (1.12), we can see that for time delay the AF has a maximum value at the commanded scan angle. This makes time delay very attractive for wide bandwidth applications and also for large arrays that have a small beamwidth and have limited bandwidth using phase delay steering.

1.3.3   Grating Lobes

The AF is a periodic function. Similar to signal processing theory, the elements in a phased array, if not sampled properly with the correct element spacing, will generate grating lobes, which are simply periodic copies of the main beam. The grating lobe locations are a function of frequency and element spacing.

Equation (1.11) can be used to derive the grating lobe locations for an ESA. The AF has maximum values when $\pi d(\frac{\sin\theta_o}{\lambda_o} - \frac{\sin\theta}{\lambda}) = \pm P\pi$, where $P = 0$, $1, 2,\dots$. Rearranging terms this can be represented as

$$\sin\theta_{GL} = \frac{\lambda}{\lambda_o}\sin\theta_o \mp P\frac{\lambda}{d}$$

(1.22)

The first term on the right-hand side of Equation (1.22) represents the location of the main beam when scanned to θ_o. The second term represents the grating lobe locations. Simplifying Equation (1.22) by setting $\lambda = \lambda_o$ and setting $\theta_o = 0°$ we find the expression

$$\sin \theta_{GL} = \mp P \frac{\lambda}{d} \tag{1.23}$$

which gives the grating lobe locations when the main beam is not scanned (at boresite). To find the element spacing required for grating lobe free scanning to 90° we set $\theta_{GL} = 90°$ and $P = 1$. This results in the following equation:

$$d = \frac{\lambda}{1 + \sin \theta_o} \tag{1.24}$$

This shows that to scan the ESA to $\theta_o = 90°$ the elements spacing must be $\frac{\lambda}{2}$. Figure 1.5 shows a plot of an ESA pattern, AF, and element pattern with element spacing of λ. The grating lobes of the AF appear at 90°, which is undesirable. However, because of pattern multiplication, the roll-off of the element pattern attenuates the grating lobe of the AF. Figure 1.6 shows

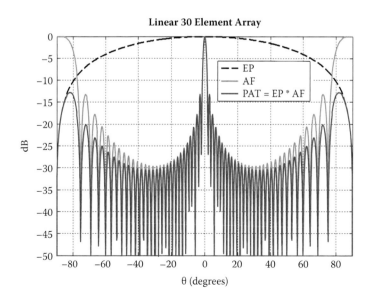

Figure 1.5 Plot of grating lobes.

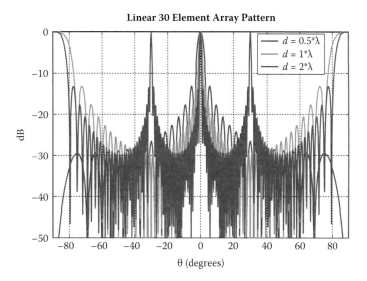

Figure 1.6 Plot of grating lobes with $d = \lambda/2$, $1^*\lambda$, and $2^*\lambda$.

pattern plots overlaid with element spacing of 2λ, λ, and $\frac{\lambda}{2}$. A good ESA design takes into account grating lobes that rob the main beam of energy. This will be discussed in the context of sine space in Chapter 2 for two-dimensional ESAs.

1.4   One-Dimensional Pattern Synthesis

Now that we have derived the pattern expression for a one-dimensional ESA, we will look at pattern plots while varying the element amplitude distribution, frequency, number of elements, and scan angle. We will begin by plotting the EP, AF, and pattern all on a single plot to visualize pattern multiplication (see Figure 1.7). The EF in this example is 1.5. The EF is dependent on the radiator design; however, 1.5 is a safe number to use for analysis. In system design, the power pattern is what determines performance and all figures will be plotted as power patterns. In Figure 1.7 all the power patterns are plotted in decibels (dB). Equation (1.25) shows how the power pattern is computed in dB.

$$F_{dB} = 10\log_{10}(EP \cdot AF)^2 = 20\log_{10} EP + 20\log_{10} AF \qquad (1.25)$$

Figure 1.8 shows the overlay of the EP, AF, and array patterns with the array scanned. The figure illustrates that the EP does not scan with the AF. The EP roll-off attenuates the AF pattern. This roll-off is governed by

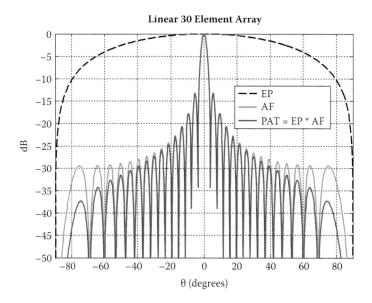

Figure 1.7 EP, AF, and pattern overlaid.

the EF. As an example, if the EF equaled 1, an additional loss of 3 dB would be added to the array pattern at $\theta = 60°$ due to the EP ($20*\log10(\cos60°) \approx -3$ dB). This difference has the most impact at large scan angles where the EP scan loss must be accounted for in analyzing performance. In addition to scan loss, the EP causes the array pattern peak to be shifted from the

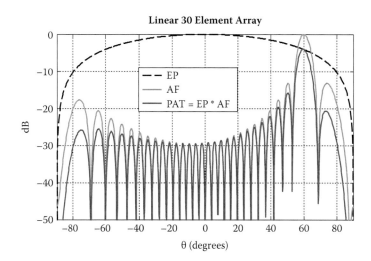

Figure 1.8 EP, AF, and pattern overlaid with scan.

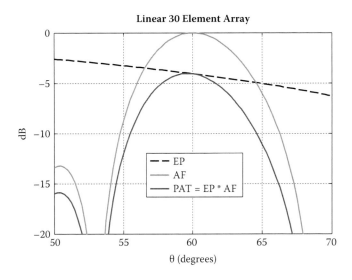

Figure 1.9 Pattern peak shifted from the scan angle due to EP roll-off.

scan angle. Figure 1.9 shows a zoomed-in view of the EP, AF, and array pattern. In the figure it is shown that the AF has its peak value at the scan angle of 60°; however, the array pattern is slightly shifted because of the EP roll-off.

1.4.1   Varying Amplitude Distribution

In Section 1.2.2, it was shown that by varying the phase of each element in the array the ESA beam can be scanned spatially. In addition to modifying the element phase, the element amplitudes (A_m) can be modified as well to lower sidelobe levels. Figure 1.10 shows the pattern for a uniform amplitude distribution across the array ($A_m = 1$). For a uniform distribution the first sidelobes are 13 dB below the peak of the main beam. When the ESA beam is scanned, the sidelobe level, relative to the main beam, changes. This is due to the roll-off of the EP impacting the array pattern. In many ESA applications it is undesirable for the main beam and the sidelobes to be close in magnitude. Signals incident on the ESA with a high enough power level could appear as signals that came in the main beam. Figure 1.11 shows an array scanned to 60°. In this example, the first sidelobe is now only 10 dB lower than the main beam.

In order to reduce the sidelobe level of the ESA, amplitude weighting can be applied. Various types of weightings can be used, similar to the filter theory; however, Taylor weighting is the most efficient aperture distribution (Taylor 1954). Figure 1.12 illustrates the difference between

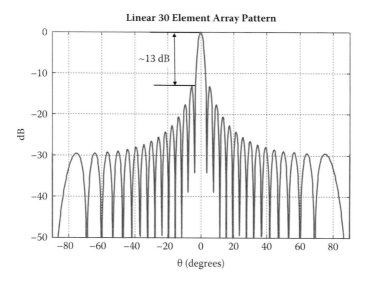

Figure 1.10 Pattern with uniform weighting.

the uniform illumination and the Taylor weighting illumination. There is a loss associated with weightings that deviate from uniform weighting that is the most efficient. This is discussed in Section 1.3.1. The pattern for the array modeled in Figure 1.10 using Taylor weighting is plotted in Figure 1.13. A 30 dB Taylor weighting is applied. The figure shows the

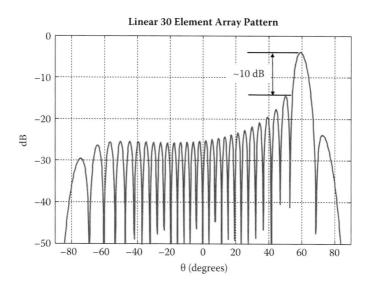

Figure 1.11 Pattern with uniform weighting scanned to 60°.

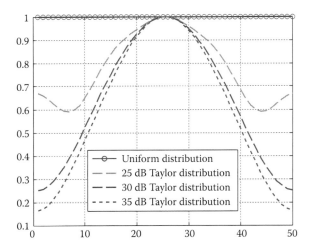

Figure 1.12 Uniform vs. Taylor weighting.

sidelobes less than 30 dB below the main beam. In practice, there are amplitude errors across the array that will cause some of the sidelobes to be above 30 dB. Typically an average sidelobe level is specified. The impacts of errors on the ESA pattern will be discussed in Chapter 2. Figure 1.14 depicts the ESA with 30 dB Taylor weighting scanned to 60°. In contrast to Figure 1.11, the sidelobes here are ~28 dB below the main beam, as opposed to 10 dB in the uniform weighting case.

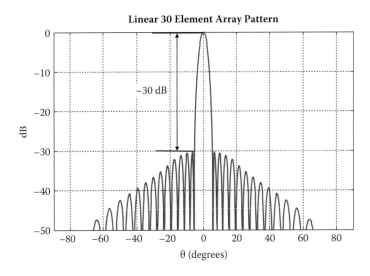

Figure 1.13 Pattern with Taylor weighting.

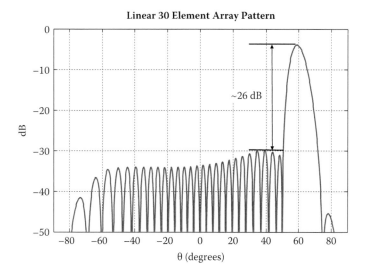

Figure 1.14 Pattern with Taylor weighting scanned to 60°.

1.4.2   Varying Frequency

As the operating frequency of the ESA is changed, the beamwidth and the sidelobes are changed as well. As will be discussed later, the beamwidth is inversely proportional to frequency. In a practical application the ESA will be required to function over an operational frequency range. As the frequency is changed over the band the beamwidth will change in size. At the high end of the band the beamwidth will be the smallest, and at the low end of the band the beamwidth will be the largest. Figure 1.15 shows a one-dimensional (1D) ESA that has an operational bandwidth of 1 GHz that spans from 2.5 GHz to 3.5 GHz. The element spacing is a half wavelength at 3.5 GHz. As discussed in Section 1.3.3, the ESA element spacing is based on the highest frequency in the operational band of interest. In Figure 1.15, the patterns at 2.5, 3.0, and 3.5 GHz are overlaid in the figure. A uniform weighting was used in order to highlight the change in beam shape and sidelobe orientation and size.

1.4.3   Varying Scan Angle

One of the great features of an ESA is the ability to electronically scan the beam. However, there is no free lunch. One of the primary impacts of electronic scan is the array pattern loss due to the EP roll-off, which has been previously discussed. However, an additional impact of electronic scan is the broadening of the main beam. As the beam is scanned

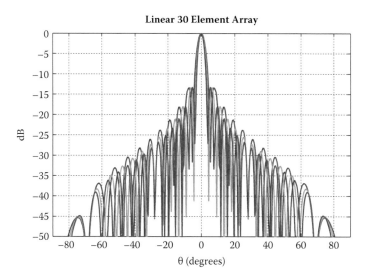

Figure 1.15 Changing beamwidth as a function of frequency.

the beam broadens at the rate of $\frac{1}{\cos\theta}$. This changes the spatial footprint of the main beam and has to be accounted for at a system level, which is beyond the scope of this book. Figure 1.16 shows an ESA pattern at several scan angles. The broadening of the main beam with increasing electronic scan is shown.

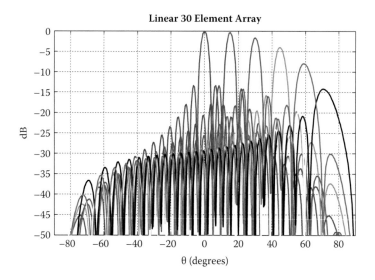

Figure 1.16 Beam broadening with electronic scan.

1.5 Conformal Arrays*

For applications involving conformal phased arrays, the normal modeling done for a planar array is no longer valid. In order to characterize the conformal array in terms of pattern distribution, sidelobe levels, and gain performance, it is important to understand how to model a nonplanar array. Since the element pattern for each element in a conformal array points in a different direction, each element contributes a different amount of energy to the main beam and sidelobes, thereby affecting the pattern.

1.5.1 Formulation

1.5.1.1 Array Pattern for a Linear Array

To understand how the pattern equation applies to a conformal array, it is best to start with a more general expression for the pattern:

$$F(\mathbf{r}) = \sum_{i}^{N} a_i EP_i(\theta, \phi) e^{jk\mathbf{r}_i \cdot \hat{\mathbf{r}}} \tag{1.26}$$

Equation (1.26) is a summation over all of the elements in the array, where N is the number of array elements. Each element has an amplitude a_i, element pattern EP_i, and phase $k\mathbf{r}_i \cdot \hat{\mathbf{r}}$, where k is the free space wave number, $\mathbf{r}_i$ is the element position vector, and $\hat{\mathbf{r}}_i$ is the spatial unit vector. It is important to note that no phase shift has been applied to the elements (i.e., no steering of the antenna beam). In order to account for beam steering, an appropriate phase term must be added to Equation (1.26). The modified pattern equation is shown below:

$$F(\mathbf{r}) = \sum_{i}^{N} a_i EP_i(\theta, \phi) e^{jk\mathbf{r}_i \cdot \hat{\mathbf{r}}} e^{-jk\mathbf{r}_i \cdot \hat{\mathbf{r}}_o} \tag{1.27}$$

In Equation (1.27), $\hat{\mathbf{r}}_o$ is the unit scan vector that corresponds to the angle in space to which the antenna beam is steered. It is important to note that the phase term, in the second exponential in Equation (1.27), represents the phase set by phase shifters in an ESA. When analyzing planar arrays, the element pattern EP_i can be moved outside of the summation in Equation (1.27). Figure 1.17 depicts a planar linear array with N elements

---

* This section is based on a technical memo written by the author and Dr. Sumati Rajan with the consultation of Dr. Daniel Boeringer at Northrop Grumman Electronic Systems (Brown and Rajan 2005).

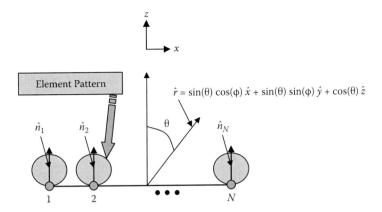

$$\hat{r} = \sin(\theta)\cos(\phi)\,\hat{x} + \sin(\theta)\sin(\phi)\,\hat{y} + \cos(\theta)\,\hat{z}$$

Figure 1.17 Unit normal vectors for the element patterns in a linear array.

to illustrate the separability of the element factor for a planar array. The unit normal for each element in Figure 1.17 points in the same direction, and the element pattern for each element is the same. This allows Equation

$$F(\mathbf{r}) = EP(\theta, \phi) \sum_{i}^{N} a_i e^{jk\mathbf{r}_i \cdot \hat{r}} e^{-jk\mathbf{r}_i \cdot \hat{r}_o} \tag{1.28}$$

Equation (1.28) can be recognized as the well-known pattern multiplication equation for an array, in which the pattern is equal to the multiplication of the element pattern and the array factor. In most array applications the element pattern (EP) is assumed to be a cosine function raised to a power called the element factor (EF). This is shown in Equation (1.29):

$$EP(\theta, \phi) = \cos^{EF/2}(\theta) \tag{1.29}$$

Equation (1.29) assumes that the elements in the array are located in the x,y plane and that the z direction is normal to the face of the array.

To better understand how the element pattern is applied in the case of a conformal array, it is helpful to express the element pattern in a different form:

$$EP(\theta, \phi) = (\hat{n} \cdot \hat{r})^{EF/2} \tag{1.30}$$

Equation (1.30) makes no assumption on what direction the array is pointed. However, when $\hat{n} = \hat{z}$, and using the definition for $(\hat{r})$, Equation (1.30) reduces to Equation (1.29). The normal vector in Equation (1.30) is normal to the surface of the array for each element.

1.5.2   Array Pattern for a Conformal Array

When modeling an array of conformal elements, Equation (1.28) is no longer valid. This is because the unit normal for each array element is oriented in a different direction, and the element pattern cannot be removed from the summation in Equation (1.27). This is depicted in Figure 1.18. For computation purposes, the normal for each element must be determined in order to properly compute the antenna pattern. Once the unit normal is known, it can be plugged into Equation (1.30) to calculate the element pattern for each individual element. Substituting Equation (1.30) into Equation (1.27) gives the following expression for the antenna pattern:

$$F(\mathbf{r}) = \sum_{i}^{N} a_i (\hat{n}_i \cdot \hat{r})^{EF/2} e^{jk\tau_i \cdot \hat{r}} e^{-jk\tau_i \cdot \hat{r}_o} \tag{1.31}$$

Equation (1.31) shows that for every angle in space (θ, ϕ), each element contributes a different amount of power to the antenna pattern. This concept is illustrated in Figure 1.18. At boresite, each element is looking through a different point in its element pattern, giving a different contribution than its neighbors.

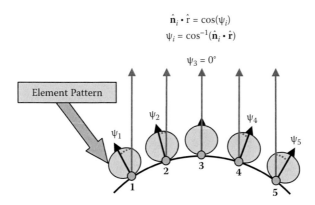

$$\hat{\mathbf{n}}_i \cdot \hat{\mathbf{r}} = \cos(\psi_i)$$
$$\psi_i = \cos^{-1}(\hat{\mathbf{n}}_i \cdot \hat{\mathbf{r}})$$

Figure 1.18 Different element pattern contributions for each individual element in a conformal array with no scan (boresite).

1.5.3   Example

1.5.3.1   Conformal One-Dimensional Array

In the following example, the array pattern for a curved linear source in the x,z plane will be calculated using Equation (1.31). The elements are assumed to lie on the arc of a circle with arbitrary radius R. Figure 1.18 depicts the example array. For simplicity, the element factor, EF, is assumed to be 1, and the amplitude weights, a_i, are uniform (i.e., $a_i = 1$). Table 1.1 shows the simplified expressions for the variables in Equation (1.31). It is important to note that the expression for the element unit normal vector, $\hat{n}_i$, in Table 1.1 is only applicable for this example geometry. For other curvatures, the expression must be modified appropriately.

Substituting the expressions in Table 1.1 into Equation (1.31) gives the following equation:

$$F(\mathbf{r}) = \sum_i^N \cos(\psi_i) e^{jk[x_i(\sin\theta - \sin\theta_o) + z(\cos\theta - \cos\theta_o)]} \qquad (1.32)$$

where $\cos(\psi_i)$ for this example is equal to

$$\cos(\psi_i) = \hat{\mathbf{n}}_i \cdot \hat{\mathbf{r}} = \frac{x_i \sin\theta + z_i \cos\theta}{\sqrt{x_i^2 + z_i^2}} \qquad (1.33)$$

Using Equation (1.33), the array pattern can then be easily computed using Equation (1.32).

Table 1.1 Variable Expressions for the Array Pattern Equation for a Curved Line Source in the x, z Plane

Variable	Simplified Expression
EF	1
a_i	1, for all i
$\hat{\mathbf{r}}$	$\sin\theta\,\hat{\mathbf{x}} + \cos\theta\,\hat{\mathbf{z}}$
$\hat{\mathbf{r}}_o$	$\sin\theta_o\,\hat{\mathbf{x}} + \cos\theta_o\,\hat{\mathbf{z}}$
$\mathbf{r}_i$	$x_i\,\hat{\mathbf{x}} + z_i\,\hat{\mathbf{z}}$
$\hat{\mathbf{n}}_i$	$\dfrac{\mathbf{r}_i}{\lvert\mathbf{r}_i\rvert}$

1.6 MATLAB Program and Function Listings

This section contains a listing of all MATLAB programs and functions used in this chapter.

1.6.1 BeamwidthCalculator.m

```
%% This Code Plots Beamwidth vs. Frequency and Scan Angle
% Arik D. Brown

%% Input Parameters
BW.k=0.886;%Beamwidth Factor (radians)
BW.f_vec=[1 5 10 15];%Frequency in GHZ
BW.lambda_vec=0.3./BW.f_vec;%meters
BW.L=1;%Aperture Length in meters
BW.thetao_vec=0:5:60;%Degrees

%% Calculate Beamwidths
[BW.lambda_mat BW.thetao_mat]=meshgrid(BW.lambda_vec,BW.
thetao_vec);
BW.mat_rad=BW.k*BW.lambda_mat./(BW.L*cosd(BW.thetao_mat));
BW.mat_deg=BW.mat_rad*180/pi;

%% Plot

figure(1),clf
plot(BW.thetao_mat,BW.mat_deg,'linewidth',2)
grid
set(gca,'fontsize',16,'fontweight','b')
xlabel('Scan Angle (Degrees)','fontsize',16,'fontweight','b')
ylabel('Beamwidth (degrees)','fontsize',16,'fontweight','b')
legend('1 GHz','5 GHz','10 GHz','15 GHz')
```

1.6.2 Compute_1D_AF.m (Function)

```
%% Function to Compute 1D AF
% Arik D. Brown

function [AF, AF_mag, AF_dB, AF_dBnorm] =...
 Compute_1D_AF(wgts,nelems,d_in,f_GHz,fo_GHz,u,uo)
lambda=11.803/f_GHz;%wavelength(in)
lambdao=11.803/fo_GHz;%wavelength at tune freq(in)

k=2*pi/lambda;%rad/in
ko=2*pi/lambdao;%rad/in

AF=zeros(1,length(u));
```

```
for ii=1:nelems
  AF = AF+wgts(ii)*exp(1j*(ii-(nelems+1)/2)*d_in*(k*u-ko*uo));
end
[AF_mag AF_dB AF_dBnorm] = process_vector(AF);
```

1.6.3   Compute_1D_EP.m (Function)

```
%% Function to Compute 1D EP
% Arik D. Brown

function [EP, EP_mag, EP_dB, EP_dBnorm] =...
  Compute_1D_EP(theta_deg,EF)

EP=zeros(size(theta_deg));

EP=(cosd(theta_deg).^(EF/2));%Volts
[EP_mag, EP_dB, EP_dBnorm] = process_vector(EP);
```

1.6.4   Compute_1D_PAT (Function)

```
%% Function to Compute 1D PAT
% Arik D. Brown

function [PAT, PAT_mag, PAT_dB, PAT_dBnorm] =...
  Compute_1D_PAT(EP,AF)

PAT=zeros(size(AF));

PAT=EP.*AF;
[PAT_mag PAT_dB PAT_dBnorm] =...
  process_vector(PAT);
```

1.6.5   process_vector.m (Function)

```
function[vectormag,vectordB,vectordBnorm] = process_
vector(vector)

vectormag=abs(vector);
vectordB=20*log10(vectormag+eps);
vectordBnorm=20*log10((vectormag+eps)/max(vectormag));
```

1.6.6   Pattern1D.m

```
% 1D Pattern Code
% Computes Element Pattern (EP), Array Factor(AF)and array
pattern (EP*AF)
% Arik D. Brown
```

```
clear all

%% Input Parameters
%ESA Parameters

%ESA opearating at tune freq
array_params.f=10;%Operating Frequency in GHz
array_params.fo=10;%Tune Frequency in GHz of the Phase
Shifter,

array_params.nelem=30;%Number of Elements
array_params.d=0.5*(11.803/array_params.fo);%Element Spacing
in Inches

array_params.EF=1.35;%EF

array_params.wgtflag=1;%0 = Uniform, 1 = Taylor Weighting

%$$$$These Parameters Only Used if array_params.wgtflag=1;
array_params.taylor.nbar=5;
array_params.taylor.SLL=30;%dB value

%Theta Angle Parameters
theta_angle.numpts=721;%Number of angle pts
theta_angle.min=-90;%degrees
theta_angle.max=90;%degrees
theta_angle.scan=0;%degrees

plotcommand.EP=0;%Plot EP if = 1
plotcommand.AF=0;%Plot EP if = 1
plotcommand.PAT=1;%Plot PAT if = 1
plotcommand.ALL=0;%Plot All patterns overlaid if = 1

%% Compute Patterns

if array_params.wgtflag==0
array_params.amp_wgts=ones(array_params.nelem,1);
else
array_params.amp_wgts=Taylor(array_params.nelem,array_
params.taylor.SLL,...
  array_params.taylor.nbar);
end

theta_angle.vec=linspace(theta_angle.min,theta_angle.max,...
  theta_angle.numpts);%degrees
theta_angle.uvec=sind(theta_angle.vec);
theta_angle.uo=sind(theta_angle.scan);
%Initialize Element Pattern, Array Factor and Pattern
```

```
array.size=size(theta_angle.vec);
array.EP=zeros(array.size);%EP
array.AF=zeros(array.size);%AF
array.PAT=zeros(array.size);

%% Compute Patterns

%Compute AF1
[array.AF, array.AF_mag, array.AF_dB, array.AF_dBnorm]=...
 Compute_1D_AF(array_params.amp_wgts,array_params.nelem,...
 array_params.d,array_params.f,array_params.fo,...
 theta_angle.uvec,theta_angle.uo);

%Compute EP
[array.EP, array.EP_mag, array.EP_dB, array.EP_dBnorm]=...
 Compute_1D_EP(theta_angle.vec,array_params.EF);

%Compute PAT
[array.PAT, array.PAT_mag, array.PAT_dB, array.PAT_dBnorm]
=...
 Compute_1D_PAT(array.EP,array.AF);

%% Plotting

if plotcommand.EP == 1
 %Plot EP in dB, Normalized
 figure,clf
 set(gcf,'DefaultLineLineWidth',2.5)
 plot(theta_angle.vec,array.EP_dBnorm,'--','color',[0 0
 0]),hold
 grid
 axis([-90 90 -50 0])
 set(gca,'FontSize',16,'FontWeight','bold')
 title(['Element Pattern'])
 xlabel('\theta (degrees)'),ylabel('dB')
end

if plotcommand.AF == 1
 %Plot PAT in dB, Normalized
 figure,clf
 set(gcf,'DefaultLineLineWidth',2.5)
 plot(theta_angle.vec,array.AF_dBnorm,'color',[0 .7 0])
 grid
 axis([-90 90 -50 0])
 set(gca,'FontSize',16,'FontWeight','bold')
 title(['Linear ',num2str(array_params.nelem),' Element
 Array Array Factor'])
 xlabel('\theta (degrees)'),ylabel('dB')
end
```

```
if plotcommand.PAT == 1
 %Plot PAT in dB, Normalized
 figure,clf
 set(gcf,'DefaultLineLineWidth',2.5)
 plot(theta_angle.vec,array.PAT_dBnorm+array.EP_
 dBnorm,'color',[0 0 1]),hold
 grid
 axis([-90 90 -50 0])
 set(gca,'FontSize',16,'FontWeight','bold')
 title(['Linear ',num2str(array_params.nelem),' Element
 Array Pattern'])
 xlabel('\theta (degrees)'),ylabel('dB')
end

if plotcommand.ALL == 1
 %Plot ALL in dB, Normalized
 figure,clf
 set(gcf,'DefaultLineLineWidth',2.5)
 plot(theta_angle.vec,array.EP_dBnorm,'--','color',[0 0
 0]),hold
 plot(theta_angle.vec,array.AF_dBnorm,'color',[0 .7 0])
 plot(theta_angle.vec,array.PAT_dBnorm+array.EP_dBnorm,'b-')
 grid
 axis([-90 90 -50 0])
% axis([50 70 -20 0])
 set(gca,'FontSize',16,'FontWeight','bold')
 title(['Linear ',num2str(array_params.nelem),' Element
 Array'])
 xlabel('\theta (degrees)'),ylabel('dB')
 legend('EP','AF','PAT = EP * AF')
end
```

1.6.7   Pattern1D_GLs.m

```
% 1D Pattern Code
% Computes Patterns for Different Element Spacing to
Illustrate Grating
% Lobes
% Arik D. Brown

clear all

%% Input Parameters
%ESA Parameters

%ESA opearating at tune freq
array_params.f=10;%Operating Frequency in GHz
array_params.fo=10;%Tune Frequency in GHz of the Phase
Shifter,
```

```
array_params.nelem=30;%Number of Elements
array_params.d1=0.5*(11.803/array_params.fo);%Element
Spacing in Inches
array_params.d2=1*(11.803/array_params.fo);%Element Spacing
in Inches
array_params.d3=2*(11.803/array_params.fo);%Element Spacing
in Inches

array_params.EF=1.35;%EF

array_params.amp_wgts=ones(array_params.nelem,1);

%Theta Angle Parameters
theta_angle.numpts=721;%Number of angle pts
theta_angle.min=-90;%degrees
theta_angle.max=90;%degrees
theta_angle.scan=0;%degrees

%% Compute Patterns

theta_angle.vec=linspace(theta_angle.min,theta_angle.max,...
 theta_angle.numpts);%degrees
theta_angle.uvec=sind(theta_angle.vec);
theta_angle.uo=sind(theta_angle.scan);

%Initialize Element Pattern, Array Factor and Pattern
array.size=size(theta_angle.vec);
array.EP=zeros(array.size);%EP

array.AF1=zeros(array.size);%AF1
array.AF2=zeros(array.size);%AF2
array.AF3=zeros(array.size);%AF3

array.PAT1=zeros(array.size);%Pattern 1
array.PAT2=zeros(array.size);%Pattern 2
array.PAT3=zeros(array.size);%Pattern 3

%% Compute Patterns

%Compute AF1
[array.AF1, array.AF1_mag, array.AF1_dB, array.AF1_dBnorm]=...
 Compute_1D_AF(array_params.amp_wgts,array_params.nelem,...
 array_params.d1,array_params.f,array_params.fo,...
 theta_angle.uvec,theta_angle.uo);
%Compute AF2
[array.AF2, array.AF2_mag, array.AF2_dB, array.AF2_dBnorm]=...
 Compute_1D_AF(array_params.amp_wgts,array_params.nelem,...
 array_params.d2,array_params.f,array_params.fo,...
 theta_angle.uvec,theta_angle.uo);
```

```
%Compute AF3
[array.AF3, array.AF3_mag, array.AF3_dB, array.AF3_dBnorm]=...
 Compute_1D_AF(array_params.amp_wgts,array_params.nelem,...
 array_params.d3,array_params.f,array_params.fo,...
 theta_angle.uvec,theta_angle.uo);

%Compute EP
[array.EP, array.EP_mag, array.EP_dB, array.EP_dBnorm]=...
 Compute_1D_EP(theta_angle.vec,array_params.EF);

%Compute PAT1
[array.PAT1, array.PAT1_mag, array.PAT1_dB, array.PAT1_
dBnorm] =...
 Compute_1D_PAT(array.EP,array.AF1);
%Compute PAT2
[array.PAT2, array.PAT2_mag, array.PAT2_dB, array.PAT2_
dBnorm] =...
 Compute_1D_PAT(array.EP,array.AF2);
%Compute PAT3
[array.PAT3, array.PAT3_mag, array.PAT3_dB, array.PAT3_
dBnorm] =...
 Compute_1D_PAT(array.EP,array.AF3);

%% Plotting

%Plot PAT in dB, Normalized
figure,clf
set(gcf,'DefaultLineLineWidth',1.5)
plot(theta_angle.vec,array.PAT1_dBnorm+array.EP_
dBnorm,'color',[0 0 1]),hold
plot(theta_angle.vec,array.PAT2_dBnorm+array.EP_
dBnorm,'color',[0 .7 0]),
plot(theta_angle.vec,array.PAT3_dBnorm+array.EP_
dBnorm,'color',[1 0 0])
grid
axis([-90 90 -50 0])
set(gca,'FontSize',16,'FontWeight','bold')
title(['Linear ',num2str(array_params.nelem),' Element
Array Pattern'])
xlabel('\theta (degrees)'),ylabel('dB')
legend('d = 0.5*\lambda','d = 1*\lambda','d = 2*\lambda')
```

1.6.8   *Pattern1D_IBW.m*

```
% 1D Pattern Code Demonstrating Beamsquint Due to IBW
Constraints
% Arik D. Brown
```

```
clear all

%% Input Parameters
%ESA Parameters
%ESA opearating at tune freq
array_params.f1=10;%Operating Frequency in GHz
array_params.deltaf=-0.2;%
array_params.f2=10+array_params.deltaf;%Operating Frequency
in GHz (Squinted Beam)
array_params.fo=10;%Tune Frequency in GHz of the Phase
Shifter,

array_params.nelem=30;%Number of Elements
array_params.d=0.5*(11.803/array_params.fo);%Element Spacing
in Inches

array_params.EF=1.35;%EF

array_params.select_wgts=0;
array_params.amp_wgts=ones(array_params.nelem,1);

%Theta Angle Parameters
theta_angle.numpts=1001;%Number of angle pts
theta_angle.min=10;%degrees
theta_angle.max=50;%degrees
theta_angle.scan=30;%degrees

%% Compute Patterns

theta_angle.vec=linspace(theta_angle.min,theta_angle.max,...
   theta_angle.numpts);%degrees
theta_angle.uvec=sind(theta_angle.vec);
theta_angle.uo=sind(theta_angle.scan);

%Initialize Element Pattern, Array Factor and Pattern
array.size=size(theta_angle.vec);
array.EP=zeros(array.size);%EP
array.AF1=zeros(array.size);%AF1 f=fo
array.AF2=zeros(array.size);%AF2 f=fo+deltaf
array.PAT=zeros(array.size);

%% Compute Patterns

%Compute AF1
[array.AF1, array.AF1_mag, array.AF1_dB, array.AF1_dBnorm]=...
 Compute_1D_AF(array_params.amp_wgts,array_params.nelem,...
 array_params.d,array_params.f1,array_params.fo,...
 theta_angle.uvec,theta_angle.uo);
```

```
%Compute AF2
[array.AF2, array.AF2_mag, array.AF2_dB, array.AF2_dBnorm]=...
 Compute_1D_AF(array_params.amp_wgts,array_params.nelem,...
 array_params.d,array_params.f2,array_params.fo,...
 theta_angle.uvec,theta_angle.uo);

%Compute EP
[array.EP, array.EP_mag, array.EP_dB, array.EP_dBnorm]=...
 Compute_1D_EP(theta_angle.vec,array_params.EF);

%Compute PAT1
[array.PAT1, array.PAT1_mag, array.PAT1_dB, array.PAT1_
dBnorm] =...
 Compute_1D_PAT(array.EP,array.AF1);

%Compute PAT2
[array.PAT2, array.PAT2_mag, array.PAT2_dB, array.PAT2_
dBnorm] =...
 Compute_1D_PAT(array.EP,array.AF2);
%% Plotting

%Plot PAT1 and PAT2 in dB, Normalized
figure 1),clf
set(gcf,'DefaultLineLineWidth',2.5)
plot(theta_angle.vec,array.PAT1_dBnorm,'k-'),hold
plot(theta_angle.vec,array.PAT2_dB - max(array.PAT1_dB),'k--
'),hold
grid
axis([25 35 -5 0])
set(gca,'FontSize',16,'FontWeight','bold')
title(['Linear ',num2str(array_params.nelem),' Element
Array'])
xlabel('\theta (degrees)'),ylabel('dB')
legend('f = f_{o}','f = f_{o} + \Delta f')
set(gca,'XTick',[25:1:35],'YTick',[-5:1:0])
```

1.6.9   Taylor.m (Function)

```
%% Code to Generate Taylor Weights
% Arik D. Brown
% Original Code Author: F. W. Hopwood

Function [wgt] = Taylor(points,sll,nbar)

r = 10^(abs(sll)/20);
a = log(r+(r*r-1)^0.5) / pi;
sigma2 = nbar^2/(a*a+(nbar-0.5)^2);
```

```
%--Compute Fm, the Fourier coefficients of the weight set
for m=1:(nbar-1)
 for n=1:(nbar-1)
  f(n,1)=1.m*m/sigma2/(a*a+(n-0.5)*(n-0.5));
 if n ~= m
  f(n,2)=1/(1.m*m/n/n);
 end
  if n==m
    f(n,2)=1;
  end
 end
 g(1,1)=f(1,1);
 g(1,2)=f(1,2);

 for n=2:(nbar-1)
  g(n,1)=g(n-1,1)*f(n,1);
  g(n,2)=g(n-1,2)*f(n,2);
 end
 F(m)=((-1)^(m+1))/2*g(nbar-1,1)*g(nbar-1,2);
end

jj = [1:points]';
xx = (jj-1+0.5)/points - 1/2; %-- column vector
W = ones(size(jj)); %-- column vector
mm = [1:nbar-1]; %-- row vector
W = W + 2*cos(2*pi*xx*mm)*F';

WPK = 1 + 2*sum(F);
wgt = W / WPK;
```

1.6.10   AmpWeightsCompare.m

```
%% Plot Different Amplitude Weights
% Arik D. Brown

%% Enter Inputs
wgts.N=50;
wgts.nbar=5;
wgts.SLL_vec=[20 30 35];

wgts.uni_vec=ones(1,wgts.N);

wgts.tay_vec=[Taylor(wgts.N,wgts.SLL_vec(1),wgts.nbar)...
 Taylor(wgts.N,wgts.SLL_vec(2),wgts.nbar)...
 Taylor(wgts.N,wgts.SLL_vec(3),wgts.nbar)];

%% Plot Weights
```

```
figure 1),clf
plot(1:wgts.N,wgts.uni_vec,'-o','linewidth',2,'color',[0 0
1]),hold
plot(1:wgts.N,wgts.tay_vec(:,1),'--
','linewidth',2.5,'color',[0 .7 0])
plot(1:wgts.N,wgts.tay_vec(:,2),'-
.','linewidth',2.5,'color',[1 0 0])
plot(1:wgts.N,wgts.tay_vec(:,3),':','linewidth',2.5,'co
lor',[.7 0 1])
grid
% xlabel('Element Number','fontweight','bold','fontsize',14)
% ylabel('Voltage','fontweight','bold','fontsize',14)
set(gca,'fontweight','bold','fontsize',14)
legend('Uniform Distribution','25 dB Taylor Distribution',...
 '30 dB Taylor Distribution','35 dB Taylor Distribution')
```

1.6.11   Pattern1D_ConformalArray.m

```
%Conformal Array Code (1D)
%Assumes the curvature is a segment of a circle centered
at (0,0)
%Array elements contained solely in the x-z plane

%% Define Parameters
%Constants
c=11.81;%Gin/s
deg2rad=pi/180;

%Paramter Def'ns
alpha=90;%deg
alpharad=alpha*deg2rad;%rad

f=10;%GHz
lambda=c/f;%inches
k=2*pi/lambda;

theta=[-90:.1:90];
thetarad=theta*deg2rad;
thetao=0;

u=sin(thetarad);w=cos(thetarad);
uo=sin(thetao*deg2rad);
wo=cos(thetao*deg2rad);

rvec=[u;w];

R=11.46*lambda;
```

```
N=36;%Number of elements
d=1.5*lambda;
alphai=-0.5*alpharad + ([1:N] -1)*alpharad/(N-1);
xi=R*sin(alphai);
zi=R*cos(alphai);

denom=sqrt(xi.^2 + zi.^2);

nveci=[xi./denom; zi./denom];

EF=1.5;
%% Compute Conformal Pattern
elempat=(nveci.'*rvec).^(0.5*EF);

cosang=acos((nveci.'*rvec))/deg2rad;
indx=find(cosang > 90);
elempat(indx)=0;

phase=k*(xi'*(u-uo) + zi'*(w-wo));
Pat=sum(elempat.*exp(1i*phase));
[Pat_mat Pat_dB Pat_dBnorm]=process_vector(Pat);
%% Plot
figure 1),clf
set(gcf,'DefaultLineLineWidth',2.5)
plot(theta,Pat_dBnorm,'-','color',[0 0 1]),grid
axis([-90 90 -50 0])
set(gca,'FontSize',16,'FontWeight','bold')
set(gca,'XTick',[-90:15:90])
title(['Conformal Array Pattern'])
xlabel('\theta (degrees)'),ylabel('dB')
```

References

Brown, Arik D., and Sumati Rajan. *Pattern Modeling for Conformal Arrays. Technical Memo.* Baltimore, MD: Northrop Grumman Electronic Systems, April 2005.

Taylor, T. T. Design of Line-Source Antennas for Narrow Beamwidth and Low Side Lobes. *IRE Transactions—Antennas and Propagation,* 1954: 16–28.

Walsh, Tom, F. W. Hopwood, Tom Harmon, and Dave Machuga. *Active Electronically Scanned Antennas (AESA).* Radar Systems Engineering Course. Baltimore, MD: Northrop Grumman Electronic Systems, December 2003.

chapter two

Electronically Scanned Array Fundamentals—Part 2

Arik D. Brown
Northrop Grumman Electronic Systems

Contents

2.1   Two-Dimensional ESA Pattern Formulation

In Chapter 1, expressions were derived for a one-dimensional (1D) ESA. The motivation for this is that a majority of the fundamental ESA concepts can be derived from the 1D expression. In practice, most ESAs are two-dimensional (2D) arrays; however, the theory expounded upon in Chapter 1 can be extended to the 2D case. Figure 2.1 shows an illustration of a 2D array of elements. The ESA antenna elements are positioned in the x-y plane and are assumed to radiate in the +z direction or the forward hemisphere. As will be discussed later, this coordinate orientation is the same as what is traditionally called antenna coordinates. Each element is assumed to possess either a phase shifter or a time delay unit to electronically scan the beam. A manifold is assumed to be behind the elements summing their individual contributions coherently.

In Chapter 1, the element spacing was represented by d. In the 2D case, two element spacing values must now be specified. The spacing in the x dimension will be denoted by dx, and dy will denote the spacing in the y dimension. The number of elements in the x direction will be represented by M (identical to Chapter 1), and N will be used to represent the number of elements in the y direction. The total number of elements can then be expressed as $M \cdot N$. After defining representations for the element spacing and number of elements in the x and y

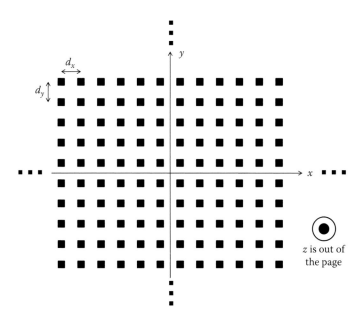

Figure 2.1 Two-dimensional ESA.

dimensions, we can now express equations for the x-y element positions in the array as

$$x_m = (m - 0.5(M+1))d_x, \text{ where } m = 1,\dots,M \tag{2.1}$$

$$y_n = (n - 0.5(N+1))d_y, \text{ where } n = 1,\dots,N \tag{2.2}$$

Equations (2.1) and (2.2) specify a rectangular grid of elements whose phase center is located at (0,0). The indexing used in Equations (2.1) and (2.2) is not unique in that the element spacing can be specified differently with a different phase center location.

The expression for a 1D AF was shown in Chapter 1 to be

$$AF = \cdot \sum_{m=1}^{M} A_m e^{j\left(\frac{2\pi}{\lambda}x_m \sin\theta\right)} \tag{2.3}$$

This must now be expanded to include the additional elements in the y dimension. The 2D AF is shown in Equation (2.4):

$$AF = \cdot \sum_{l=1}^{M \cdot N} C_l e^{j\left(\frac{2\pi}{\lambda}x_l \sin\theta\cos\phi + \frac{2\pi}{\lambda}y_l \sin\theta\sin\phi\right)} \tag{2.4}$$

where C_l is a complex voltage that can be represented as $C_l = c_l e^{j\Theta_l}$. Setting $\Theta_l = (\frac{2\pi}{\lambda}x_l \sin\theta_o \cos\phi_o + \frac{2\pi}{\lambda}y_l \sin\theta_o \sin\phi_o)$, Equation (2.4) can then be expressed as

$$AF = \sum_{l=1}^{M \cdot N} c_l e^{j\left[\left(\frac{2\pi}{\lambda}x_l \sin\theta\cos\phi + \frac{2\pi}{\lambda}y_l \sin\theta\sin\phi\right) - \left(\frac{2\pi}{\lambda}x_l \sin\theta_o \cos\phi_o - \frac{2\pi}{\lambda}y_l \sin\theta_o \sin\phi_o\right)\right]} \tag{2.5}$$

Rearranging terms in Equation (2.5) and substituting $a_l \cdot b_l$ for c_l yields

$$AF = \sum_{l=1}^{M \cdot N} a_l e^{j\left(\frac{2\pi}{\lambda}x_l \sin\theta\cos\phi - \frac{2\pi}{\lambda}x_l \sin\theta_o \cos\phi_o\right)} \cdot b_l e^{j\left(\frac{2\pi}{\lambda}y_l \sin\theta\sin\phi - \frac{2\pi}{\lambda}y_l \sin\theta_o \sin\phi_o\right)} \tag{2.6}$$

If we assume that for each row n the a_l are constant and that for each row m the b_l are constant, Equation (2.6) can be written as

$$AF = \sum_{m=1}^{M} a_m e^{j\left(\frac{2\pi}{\lambda}x_m \sin\theta\cos\phi - \frac{2\pi}{\lambda}x_m \sin\theta_o \cos\phi_o\right)} \cdot \sum_{n=1}^{N} b_n e^{j\left(\frac{2\pi}{\lambda}y_n \sin\theta\sin\phi - \frac{2\pi}{\lambda}y_n \sin\theta_o \sin\phi_o\right)} \tag{2.7}$$

This condition is defined as separable weights, which means the 2D AF can be calculated by multiplying the 1D AFs for x and y. For weightings that are not separable, such as circular weighting, Equation (2.7) *cannot* be used and Equation (2.5) should be used. Using Equation (2.5), the total 2D array pattern is

$$F(\theta,\phi) = \cos^{\frac{EF}{2}} \theta \cdot \sum_{l=1}^{M \cdot N} c_l e^{j\left[\left(\frac{2\pi}{\lambda}x_l \sin\theta\cos\phi + \frac{2\pi}{\lambda}y_l \sin\theta\sin\phi\right) - \left(\frac{2\pi}{\lambda}x_l \sin\theta_o \cos\phi_o - \frac{2\pi}{\lambda}y_l \sin\theta_o \sin\phi_o\right)\right]}$$

(2.8)

2.2   ESA Spatial Coordinate Definitions

When computing the spatial pattern for an ESA it is important to delineate what coordinate system is being used. Depending on the application, some coordinate systems may be more advantageous than others. Figure 2.2 depicts a 2D ESA in three-dimensional space. For convenience the ESA is positioned in the x–y plane, and it is radiating in the +z direction. The +z direction is referred to as the forward hemisphere, and the –z direction is referred to as the backward hemisphere. The point R, as shown in Figure 2.2, represents a point in space whose origin is at (0,0,0) and coincides with the boresite position of the ESA. The dashed lines, which reside in the planes of x–y, y–z, and x–z, are the projections of point R in those planes and are represented as P_{ij} for $i,j = x, y,$ or z. This picture will serve as the basis for understanding the other coordinate systems described in this section. In the following coordinate systems discussed, the purpose is to represent each point in space with a corresponding angle pair. These angles are then used to describe the spatial distribution of the ESA pattern, which correspondingly relates to performance.

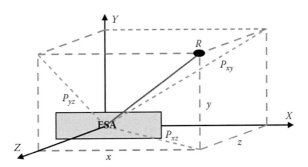

Figure 2.2 Two-dimensional ESA in three-dimensional space. (From Batzel, U., et al., *Angular Coordinate Definitions*, Antenna Fundamentals Class, Northrop Grumman Electronic Systems, Baltimore, MD, August 2010.)

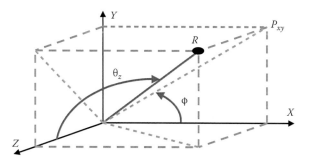

Figure 2.3 Antenna coordinates. (From Batzel, U., et al., *Angular Coordinate Definitions*, Antenna Fundamentals Class, Northrop Grumman Electronic Systems, Baltimore, MD, August 2010.)

2.2.1 Antenna Coordinates

Figure 2.3 depicts what is commonly called antenna coordinates. In this coordinate system each point R in space is represented by the angles θ_z and ϕ. θ_z is the angle subtended from the z axis to the point R. The angle ϕ is the angle between the projection of R onto the x–y axis and the x axis. This coordinate system definition is very intuitive, as it is simply the spherical coordinate system. Correspondingly, the point R can be represented as $R = (\sin\theta_z \cos\phi, \sin\theta_z \sin\phi, \cos\theta_z)$ when the magnitude of R is set to unity.

As an example, if an ESA's main beam is scanned in elevation to 45°, there is a corresponding θ_z and ϕ that are associated with that point in space. For this example, this would correspond to $\theta_z = 45°$, $\phi = 90°$ as shown in Figure 2.4. Figure 2.5 shows three different scan conditions when $\theta_z = 45°$. The azimuth scan refers to scanning the beam in the x–z plane ($\phi = 0°$).

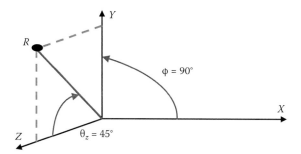

Figure 2.4 Antenna coordinates: Elevation scan to 45°. (From Batzel, U., et al., *Angular Coordinate Definitions*, Antenna Fundamentals Class, Northrop Grumman Electronic Systems, Baltimore, MD, August 2010.)

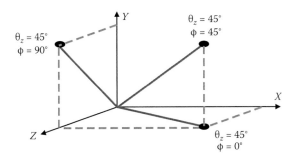

Figure 2.5 Azimuth scan, diagonal scan, and elevation scan.

The intercardinal (diagonal) scan refers to scanning the beam by setting $\phi = 45°$. Finally, the elevation scan is the same as was described in Figure 2.4 ($\phi = 90°$). Table 2.1 lists the values of θ_z and ϕ for the three different scan cases. When θ_z is kept constant, this traces a cone out in space whose apex is at $z = 0$ and whose base traces out a circle parallel to the x–y plane. This is typically called a scan cone angle. If we were to rotate Figure 2.5 about the z axis and view it looking perpendicular to the x–y plane the lines would trace out a circle.

2.2.2   Radar Coordinates

Figure 2.6 graphically defines the spatial angles used for radar coordinates. Similar to antenna coordinates, two angles are required to define a point in three-dimensional space. The two angles are θ_{AZ} and θ_{EL}. θ_{AZ} is defined as the angle subtended by the projection of R onto the x–z plane with the z axis. θ_{EL} is defined as the angle subtended by the vector to point R and the x–z plane. We'll use the same example as in the previous section and observe what values of θ_{AZ} and θ_{EL} correspond to a 45° scan in elevation. Figure 2.7 shows that for this scan case, $\theta_{AZ} = 0°$ and $\theta_{EL} = 45°$.

From a radar perspective, this coordinate system is more intuitive than that of antenna coordinates. In a radar system, the ESA main beam is typically scanned in some type of raster fashion where the beams are distributed spatially in rows and columns. Figure 2.8 shows an exemplar

Table 2.1 Angle Values for Various Scan Types

Scan Type	θ_z	ϕ
Azimuth scan	$\theta_z = 0$ to $90°$	$\phi = 0°$, $180°$
Intercardinal scan	$\theta_z = 0$ to $90°$	$\phi = 45°$, $225°$
Elevation scan	$\theta_z = 0$ to $90°$	$\phi = 90°$, $270°$

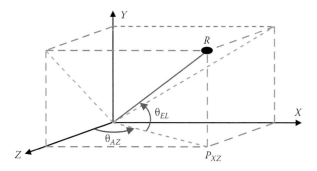

Figure 2.6 Radar coordinates. (From Batzel, U., et al., *Angular Coordinate Definitions*, Antenna Fundamentals Class, Northrop Grumman Electronic Systems, Baltimore, MD, August 2010.)

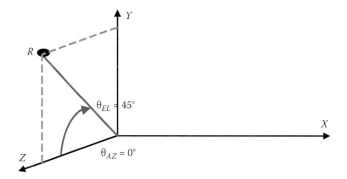

Figure 2.7 Radar coordinates: elevation scan to 45°.

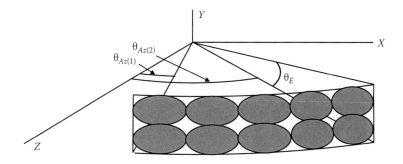

Figure 2.8 Raster scan of ESA beams. (From Batzel, U., et al., *Angular Coordinate Definitions*, Antenna Fundamentals Class, Northrop Grumman Electronic Systems, Baltimore, MD, August 2010.)

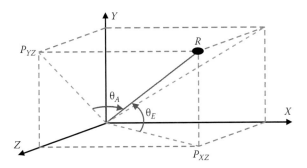

Figure 2.9 Antenna cone angle coordinates. (From Batzel, U., et al., *Angular Coordinate Definitions*, Antenna Fundamentals Class, Northrop Grumman Electronic Systems, Baltimore, MD, August 2010.)

of a raster scan of a beam in azimuth and elevation. In the figure each row corresponds to an azimuth scan where θ_{EL} is constant and θ_{AZ} is varied. This type of application lends itself to the radar coordinate system.

2.2.3   Antenna Cone Angle Coordinates

Figure 2.9 shows a pictorial definition of antenna cone angle coordinates. The two angles that specify a point in space for this coordinate system are θ_A and θ_E. θ_A is defined as the angle subtended by the projection of R onto the y–z plane and R. θ_E is defined as the angle between the point R and the x–z plane. From this definition we see that $\theta_E = \theta_{EL}$ (radar coordinates).

Any of the coordinate systems previously mentioned can be used to calculate spatial coordinates for an ESA. In a real system application, an antenna engineer may be using a coordinate system different than the system engineer. Because of this, it is good to have the angular transformations between coordinates systems so that there is consistency in requirements flowdown and system performance evaluation. Tables 2.2

Table 2.2 Angle Transformation Given Antenna Coordinate System Angles

		Given antenna angles θ_z and ϕ
Radar coordinates	θ_{AZ}	$\mathrm{atan2}(\sin(\theta_z)\cdot\cos(\phi),\cos(\theta_z))$
	θ_{EL}	$\mathrm{asin}(\sin(\theta_z)\cdot\sin(\phi))$
Antenna cone angles	θ_A	$\mathrm{asin}(\sin(\theta_z)\cdot\cos(\phi))$
	θ_E	$\mathrm{asin}(\sin(\theta_z)\cdot\sin(\phi))$

Table 2.3 Angle Transformation Given Radar Coordinate System Angles

		Given radar angles θ_{AZ} and θ_{EL}
Antenna angles	θ_Z	$\text{acos}(\cos(\theta_{Az}) \cdot \cos(\theta_{EL}))$
	ϕ	$\text{atan2}(\sin(\theta_{EL}), \sin(\theta_{Az}) \cdot \cos(\theta_{EL}))$
Antenna cone angles	θ_A	$\text{asin}(\sin(\theta_{Az}) \cdot \cos(\theta_E))$
	θ_E	θ_{EL}

to 2.4 provide a summary of the angular transformations for the antenna, radar, and cone angle coordinate systems.

2.3 Sine Space Representation

An alternative to using angular coordinates for modeling ESAs is the sine space representation. Sine space is simply a hemispherical projection of three-dimensional space onto a two-dimensional surface. Figure 2.10 graphically illustrates how three-dimensional space is mapped into two-dimensional space. Sine space is represented by the following variables: u, v, and w. Although the sine space variables can be computed using any of the three angular coordinate systems previously discussed, antenna coordinates provide a very intuitive comparison and will be elaborated on in the following discussion. The conversion to sine space from the other coordinates systems is shown in Table 2.5.

The expressions for sine space are

$$u = \sin\theta_Z \cos\phi \tag{2.9}$$

$$v = \sin\theta_Z \sin\phi \tag{2.10}$$

$$w = \cos\theta_Z \tag{2.11}$$

Table 2.4 Angle Transformation Given Antenna Cone Angle Coordinate System Angles

		Given antenna cone angles θ_A and θ_E
Antenna angles	θ_Z	$\text{asin}\left(\sqrt{\sin^2(\theta_A) - \sin^2(\theta_E)}\right)$
	ϕ	$\text{atan2}(\sin(\theta_E), \sin(\theta_A))$
Radar angles	θ_{AZ}	$\text{asin}\left(\dfrac{\sin(\theta_A)}{\cos(\theta_E)}\right)$
	θ_{EL}	θ_E

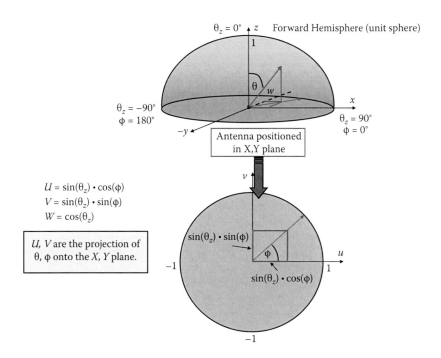

Figure 2.10 Sine space representation. (From Long, J., and Schmidt, K., *Presenting Antenna Patterns with FFT and DFT*, Antenna Fundamentals Class, Northrop Grumman Electronic Systems, Baltimore, MD, July 2010.)

These expressions are the traditional expressions for x, y, and z in spherical coordinates. Using Equations (2.9) to (2.11), a simplified form for the 2D AF can be written as

$$AF = \sum_{l=1}^{M \cdot N} c_l e^{j\left[\left(\frac{2\pi}{\lambda}x_l u + \frac{2\pi}{\lambda}y_l v\right) - \left(\frac{2\pi}{\lambda}x_l u_o - \frac{2\pi}{\lambda}y_l v_o\right)\right]} \qquad (2.12)$$

Table 2.5 Conversion to Sine Space from Angular Coordinates

Sine Space	Antenna (θ_z, ϕ)	Radar $(\theta_{AZ}, \theta_{EL})$	Antenna Cone (θ_A, θ_E)
u	$\sin\theta_z \cos\phi$	$\sin\theta_{AZ} \cos\theta_{EL}$	$\sin\theta_A$
v	$\sin\theta_z \sin\phi$	$\sin\theta_{EL}$	$\sin\theta_E$
w	$\cos\theta_z$	$\cos\theta_{AZ} \cos\theta_{EL}$	$\cos\left(\mathrm{asin}\left(\frac{\sin(\theta_A)}{\cos(\theta_E)}\right)\right)\cos\theta_E$

Several characteristics of the AF in sine space are:

- Constant beamwidth is independent of scan.
- Peak of the scanned beam in sine space is a distance of $\sin\theta_z$ from (0,0).
- For the forward hemisphere, u and v vary from −1 to +1 and w varies from 0 to 1.

For a planar array, w does not contribute as there is no z component in the exponent of Equation (2.12). For 2D nonplanar arrays the w term must be included.

2.4   ESA Element Grid

An ESA is composed of antenna elements that are arranged according to a defined element spacing. In Chapter 1, we introduced a linearly spaced 1D array of elements. It was shown that grating lobes arise due to the periodic nature of the AF and are a function of the element spacing. For a 2D configuration, the same relationships hold between grating lobes and element spacing, with the difference being there are grating lobes in both the x and y dimensions spatially. An alternative to a rectangular grid of elements is a triangular grid. The triangular grid has unique properties that will be elaborated on further in this chapter.

2.4.1   Rectangular Grid

Figure 2.1 illustrates a rectangular grid of elements with a linear spacing in both the x and y dimensions. Similar to the 1D case, now an expression is required to determine the grating lobes that occur due to the arrangement of elements in the y dimension. The equations for the grating lobes can be represented as (Mailloux 1994):

$$u_m = u_o + m\frac{\lambda}{d_x}, \qquad m = 0,\ \pm 1,\ \pm 2,...$$

$$v_n = v_o + n\frac{\lambda}{d_y}, \qquad n = 0,\ \pm 1,\ \pm 2,...$$

(2.13)

with the following relations:

$$\cos\theta_{mn} = \left(1 - u_m^2 - v_n^2\right)^{\frac{1}{2}}$$

(2.14)

Figure 2.11 shows the grating lobe locations in sine space for a 2D rectangular grid with half-wavelength spacing for both d_x and d_y.

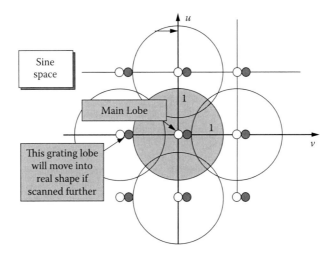

Figure 2.11 Grating lobe diagram for a rectangular grid with half-*l* spacing. (From Davis, D., *Grating Lobes Analysis*, Antenna Fundamentals Class, Northrop Grumman Electronic Systems, Baltimore, MD, July 2010.)

Half-wavelength element spacing has been chosen to simplify the grating lobe expressions in Equation (2.13). The grating lobe locations (u_m, v_n) are then integers and can be easily plotted, as shown in Figure 2.11. In Figure 2.11, the region inside the unit circle is referred to as visible space and is shaded. This is because the constant circle of radius 1 corresponds to a constant circle of $\theta = 90°$, which is the extent of the forward hemisphere. In Equation (2.13), u_o corresponds to the main beam and for boresite ($\theta_o = 0°$, $\phi_o = 0°$) has a value of 0. What is readily apparent is that as the main beam is electronically scanned, the grating lobe locations move with the main beam with a fixed offset that is proportional to an integer multiple of $\frac{\lambda}{d}$. This is demonstrated in Figures 2.12 to 2.14. Each figure shows the grating lobe locations for a boresite condition with the overlaid position of the ESA beam and its corresponding grating lobes. The element spacings shown are for $d_x = d_y = \frac{\lambda}{4}$, $\frac{\lambda}{2}$, and λ. For all cases, when the main beam is electronically scanned the grating lobes move with the main beam. If the element spacing is greater than half wavelength, for certain angles grating lobes will be present in visible space, which is undesired and unacceptable for most applications. In contrast, if the element spacings are made less than half wavelength, more margin is added, but at a price. More electronics are needed for the increased number of elements, which translates to a more expensive ESA.

The region in visible space where the main beam can be scanned is referred to as the grating lobe free scan volume. This can simply be illustrated by drawing unit circles around each of the grating lobes. Wherever

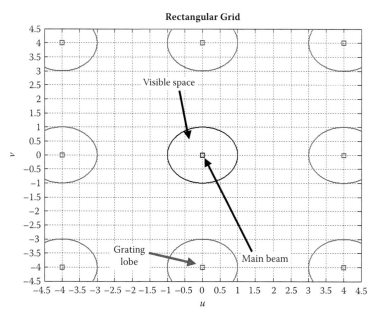

Figure 2.12 Grating lobe diagram for a rectangular grid for boresite and electronic scan ($d_x = d_y = \frac{\lambda}{2}$).

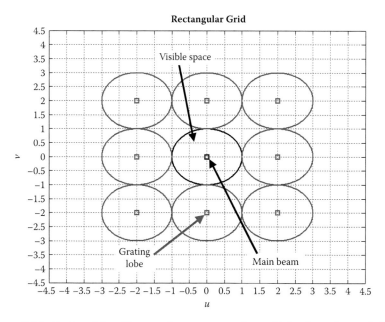

Figure 2.13 Grating lobe diagram for a rectangular grid for boresite and electronic scan ($d_x = d_y = \frac{\lambda}{4}$).

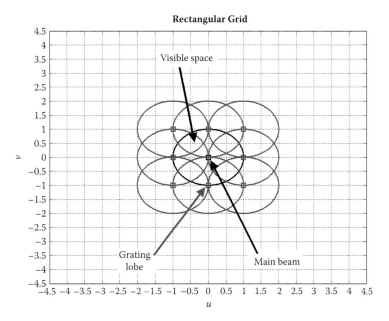

Figure 2.14 Grating lobe diagram for a rectangular grid for boresite and elec-tronic scan ($d_x = d_y = \lambda$).

the grating lobe circles do not intersect visible space corresponds to the grating lobe free scan volume. Figures 2.15 and 2.16 illustrate the depen-dence of the grating lobe free scan volume on the element spacing. In Figure 2.15, the grating lobe free scan volume is the entire unit circle or all of visible space. The ESA beam can be scanned electronically anywhere in visible space without the presence of grating lobes. In Figure 2.16 this is not the case. The grating lobe free scan volume for the larger element spacing is now limited to a portion of visible space. For scan angles cor-responding to the overlapped unit circles of the main beam and its asso-ciated grating lobes, grating lobes will appear in real space. Only in the regions where there is no overlap of the unit circles can the ESA beam be scanned without grating lobes. This is denoted by the shaded region in Figure 2.16. An additional grating lobe characteristic is that in the inter-cardinal (diagonal) plane the ESA can be scanned farther. This is because the diagonal distance of the main beam from the diagonal grating lobes is $\sqrt{2} \cdot \frac{\lambda}{d}$. For applications where the grating lobe free scan volume in Figure 2.16 would be acceptable, a cost savings would be realized in elec-tronics because half-wavelength spacing is not required, which means a reduced element count.

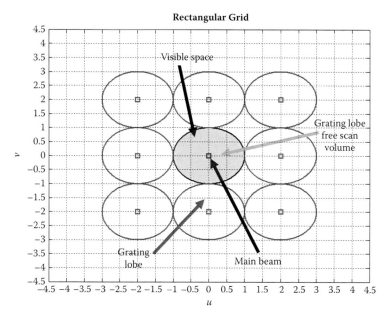

Figure 2.15 Grating lobe free scan volume for a max scan of 60° in both azimuth and elevation for $d_x = d_y = \frac{\lambda}{2}$.

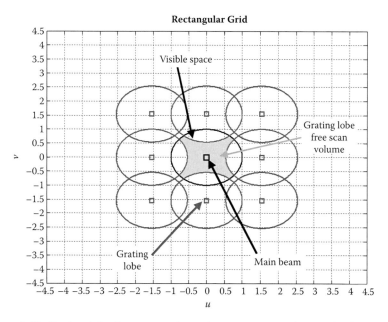

Figure 2.16 Grating lobe free scan volume for a max scan of 60° in both azimuth and elevation for $d_x = d_y = \frac{\lambda}{1.866}$.

2.4.2   Triangular Grid

Figure 2.17 shows a triangular grid of elements. In the previous section, it was mentioned that a way to save cost is by reducing the number of elements using element spacing greater than a half wavelength. Triangular grids provide another way to reduce element count while maintaining scan performance. For a rectangular grid the area per element is $d_x \cdot d_y$. The area per element for a triangular grid $2d_x \cdot d_y$ is. So for a fixed aperture size, less elements are required for a triangular grid. Furthermore, it can be shown that for the same amount of grating lobe suppression, a rectangular grid requires 16% more elements than a triangular grid (Skolnik 1990).

Using the element spacing definitions as shown in Figure 2.17, the expressions for the grating lobes are (Skolnik 1990):

$$u_m = u_o + m\frac{\lambda}{2d_x}, \quad v_n = v_o + n\frac{\lambda}{2d_y}$$

$$m, \ n = 0, \ \pm1, \ \pm2,...$$

$$m + n \text{ is even}$$

(2.15)

The derivation for these expressions are included in Appendix 3. Using the expressions in Equation (2.15), grating lobe free scan volume plots can be

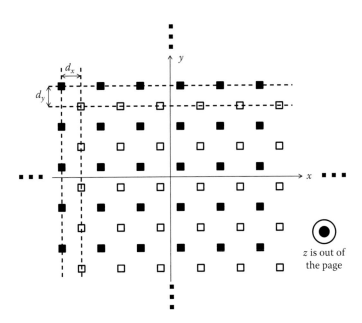

Figure 2.17 Triangular grid of array elements.

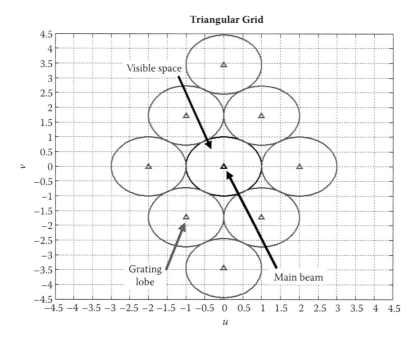

Figure 2.18 Grating lobe free scan volume for a max scan of 60° in both azimuth and elevation for $d_x = d_y = \frac{\lambda}{1.866}$ for a triangular grid.

shown similar to those in the previous section. Figure 2.18 shows the grating lobe free scan volume for a triangular grid. In Figure 2.18 we see that in sine space the grating lobes are oriented in a triangular manner just as the element grid. This can be advantageous in applications where the triangular grid provides a better scan volume due to the orientation of the grating lobes.

2.5 Two-Dimensional Pattern Synthesis Analysis

2.5.1 Pattern Synthesis

The expression for a 2D ESA array factor was shown to be

$$AF = \sum_{l=1}^{M \cdot N} c_i e^{j\left[\left(\frac{2\pi}{\lambda}x_l u + \frac{2\pi}{\lambda}y_l v\right) - \left(\frac{2\pi}{\lambda}x_l u_o - \frac{2\pi}{\lambda}y_l v_o\right)\right]} \tag{2.16}$$

The complete expression for the 2D ESA pattern is therefore

$$F(\theta, \varphi) = \cos^{\frac{EF}{2}}\theta \cdot \sum_{l=1}^{M \cdot N} c_i e^{j\left[\left(\frac{2\pi}{\lambda}x_l u + \frac{2\pi}{\lambda}y_l v\right) - \left(\frac{2\pi}{\lambda}x_l u_o - \frac{2\pi}{\lambda}y_l v_o\right)\right]} \tag{2.17}$$

Table 2.6 ESA Parameter Values Used for the 2D Pattern Plots in Section 2.5.1

Parameter	Value
Frequency	$f = f_o = 8\ \text{GHz}$
Number of elements	$n_x = 30,\ n_y = 30,\ n = n_x \cdot n_y = 900$
Element spacing	$d_x = d_y = \dfrac{\lambda}{1 + \sin(60°)} = 0.536 \cdot \lambda$
Element factor	$EF = 1.5$

The remainder of this chapter will illustrate 2D array patterns plotted in sine space for convenience. The angular equivalent plots can easily be generated by using the conversions from sine space to angle space shown in Table 2.5. The following three subsections will show patterns for an ESA using the parameters contained in Table 2.6. An element spacing less than 0.5 λ is used to show grating lobes effects for electronic scan beyond 60° for this set of ESA parameters.

2.5.1.1   Ideal Patterns

Ideal patterns using Equation (2.17) are useful to provide a high level of confidence in the ESA's scan performance. Although it is necessary to include the impacts of errors, using ideal patterns provides an initial basis with which to start a design. Amplitude and phase errors perturb the ideal pattern and primarily impact sidelobe levels. A great benefit of ESAs that have a large number of elements is that the main beam remains relatively unaltered. The main beam of an ESA is well behaved even in the presence of errors.

Figures 2.19 and 2.20 show boresite antenna patterns in radar coordinates and sine space, respectively. The remaining plots will be shown in sine space. The plots in Figure 2.21 show the antenna pattern as a function of electronic scan in the principal planes and in the intercardinal (diagonal) plane. In the 2D plots, the grating lobe can be seen coming into real space. To avoid this in practice, margin can be added to the element spacing equations shown in Chapter 1. Figure 2.22 shows the array pattern for electronic scan beyond 60°. The element spacing chosen for this example does not support a grating lobe free scan volume for scan angles greater than 60°. Thus, the grating lobes appear in real space, which is undesired.

Figure 2.23 highlights the reduction in SLL with uniform illumination and a 30 dB Taylor illumination. In applications where low SLLs are required, amplitude weighting is a valuable design tool. When amplitude

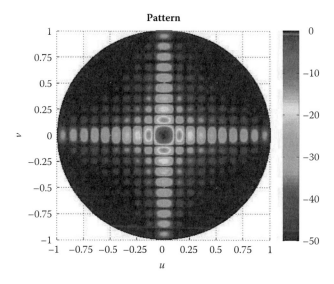

Figure 2.19 Boresite antenna pattern (no electronic scan) in radar coordinates. See insert.

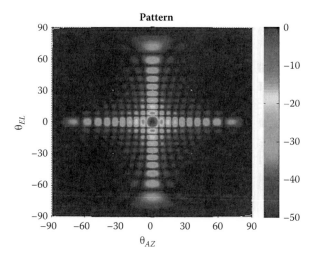

Figure 2.20 Boresite antenna pattern (no electronic scan) in sine space. See insert.

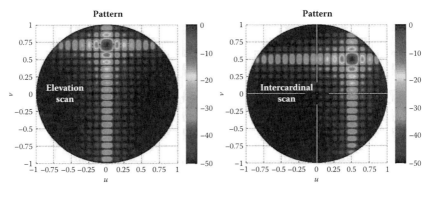

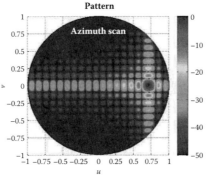

Figure 2.21 Electronically scanned antenna patterns in the principal planes ($\theta_o = 60°$, $\phi_o = 0°$ and $\theta_o = 60°$, $\phi_o = 90°$) and intercardinal plane ($\theta_o = 60°$, $\phi_o = 45°$). See insert.

weighting is implemented there is no free lunch. A reduction in SLLs comes at the price of increased beamwidth and reduced gain. The reduction in gain is called the taper loss (Mailloux 1994) and can be represented by

$$TL = \frac{\left|\sum c_l\right|^2}{n \sum \left|c_l\right|^2} \tag{2.18}$$

where n is the number of elements and c are the amplitude weights shown in Equation (2.17).

2.5.1.2 Error Effects

Many errors are introduced in an actual ESA design. These errors are both random and correlated and are caused by imperfect components and signal distribution networks (Mailloux 1994). If these errors are not controlled, they can result in large sidelobe levels, which is undesired. The primary

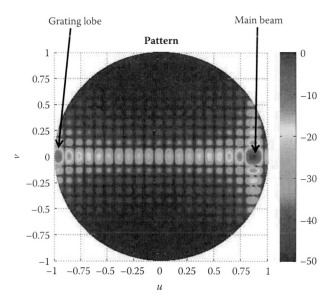

Figure 2.22 Electronic scan beyond 60° allows the fully formed grating lobe to appear in real space. See insert.

source of correlated errors is due to quantization of the phase shifters and attenuators used to implement electronic scan and amplitude taper in an ESA. An additional source of correlated error is time delay implemented at the subarray level for large wideband ESAs (Mailloux 1994). The latter will be covered in Chapter 3. Random amplitude and phase errors in an ESA can be attributed to things such as failed elements, manufacturing tolerances, etc. In order to achieve good sidelobe level performance both quantization and random errors should be simulated.

2.5.1.2.1 Quantization Effects. In an ESA, phase shifters (or time delay units) provide the progressive phase across the array required to scan the beam. A phase shifter is usually characterized by its number of bits (N). The least significant bit (LSB) is then calculated as

$$LSB = \frac{360°}{2^N} \tag{2.19}$$

By quantizing the phase at each element the phase across the array is implemented as a staircase approximation to the required phase. This is illustrated in Figure 2.24. This produces a periodic triangular phase error that leads to undesired sidelobes with a grating lobe like periodicity

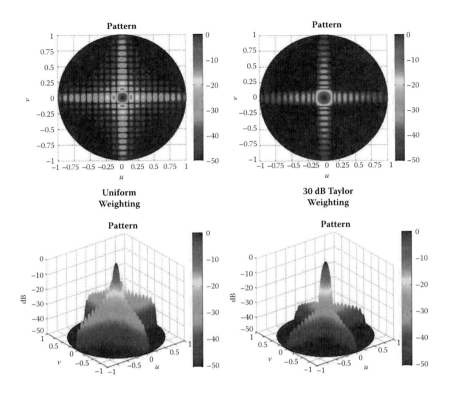

Figure 2.23 SLL reduction using a 30 dB Taylor weighting compared to a uniform distribution. See insert.

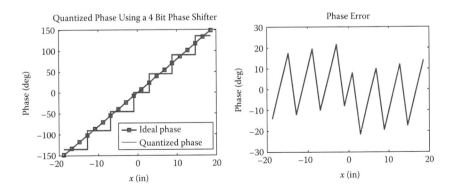

Figure 2.24 Staircase approximation for a 4-bit phase shifter.

(Mailloux 1994). Miller (1964) derived expressions for the peak and average SLL due to quantization error only, which are

$$AverageSLL = \frac{1}{3n_{elem}\varepsilon} \frac{\pi^2}{2^{2N}}$$

$$PeakSLL = \frac{1}{2^{2N}}$$

(2.20)

The quantization of amplitude taper has an effect similar to that mentioned with phase. Instead of a smooth aperture amplitude distribution, the amplitude is quantized by N bit attenuators. This appears as another source of error.

Figure 2.25 shows the impacts of quantization. The pattern with 2-bit phase shifters and attenuators has significant sidelobes due to quantization error, compared to the ideal pattern. Figure 2.26 shows the comparison of an ideal pattern with that of the pattern employing 6-bit phase

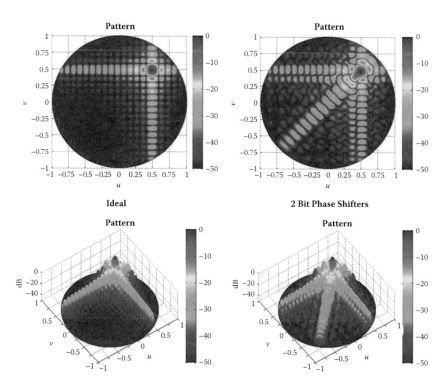

Figure 2.25 Pattern from Figure 2.20. Boresite antenna pattern (no electronic scan) in sine space with 2-bit phase shifters and attenuators and a pattern with 6-bit phase shifters and attenuators. See insert.

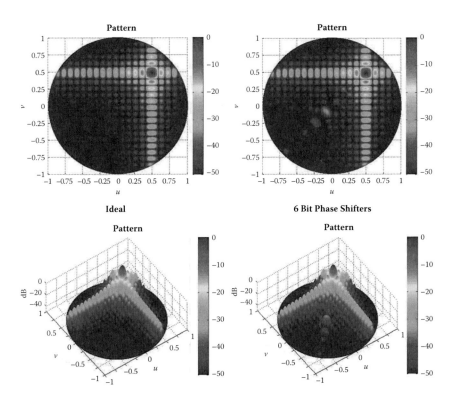

Figure 2.26 Comparison of an ideal pattern and a pattern with 6-bit phase shifters and attenuators. See insert.

shifters and attenuators. The pattern with 6-bit element control has significant performance improvement over the pattern with only 2 bits. The quantization error raises the average SLL but not enough for significant SLL impact employing 6-bit phase shifters and attenuators.

2.5.1.2.2   Random Error Effects (Amplitude and Phase).   Random errors that occur, such as failed elements, can be modeled as having a Gaussian distribution with zero mean and variance for both phase and amplitude (Mailloux 1994). The average SLL due to random errors can be expressed as (Mailloux 1994)

$$\overline{\sigma^2} = \frac{(\pi)^{\frac{1}{2}}\overline{\epsilon^2}}{D_A^{\frac{1}{2}}P} \tag{2.21}$$

where $\overline{\sigma^2}$ is the average SLL, DA is the directive gain, P is the probability of an element working (not failed), and $\overline{\epsilon^2}$ is the error variance, which is a function of the phase and amplitude errors that have a Gaussian

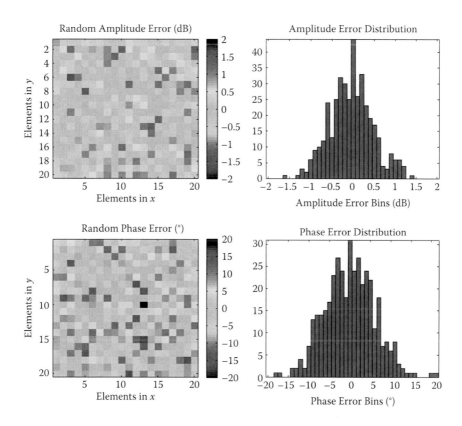

Figure 2.27 Plot of Gaussian distributed random phase and amplitude errors with a variance of 6° and 0.5 dB, respectively.

distribution. Figure 2.27 shows a plot of 6° phase and 0.5 dB amplitude random errors (1σ). The phase and amplitude error values are typically specified to a performance requirement. Figure 2.28 illustrates the impacts of the random errors on the pattern as compared to that of an ideal pattern.

2.5.2   Tilted ESA Patterns

Up to this point, we have assumed that the platform on which the ESA is installed is co-boresited with the ESA. In many installations this is not the case. The ESA may be installed in such a way that the normal of the ESA does not coincide with the normal of the platform. In Section 2.2, antenna coordinate systems were defined. In many instances, the boresite direction of the radar coordinates does not coincide with that of the ESA. As an example, radar ESAs for ship installations may be tilted upward to provide optimal coverage for the area above the horizon (the ocean).

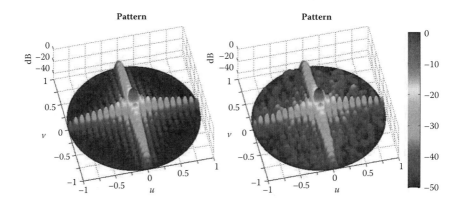

Figure 2.28 Comparison of an ideal pattern with a pattern that has the random amplitude and phase errors shown in Figure 2.27. See insert.

Additionally, on platforms that are subject to motion, such as a ship in rough seas, the ESA may be subject to motion about the x, y, z axis, which will change how the pattern is distributed spatially (Konapelsky 2002).

Figure 2.29 shows a side view of an ESA tilted relative to the x axis. In Figure 2.29 the x axis is normal to the page. Different definitions can be used to describe rotation about the three principal axes. In this book we will use the following definitions. A pitch refers to a rotation about the x axis, a roll refers to rotation about the z axis (normal to the face of the array), and a yaw refers to a rotation about the y axis. Figure 2.30 pictorially shows roll, pitch, and yaw as previously defined.

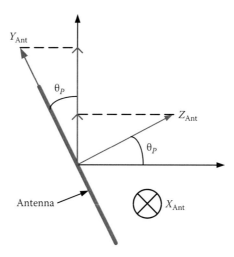

Figure 2.29 Example of an ESA tilted relative to the boresite of radar system coordinates.

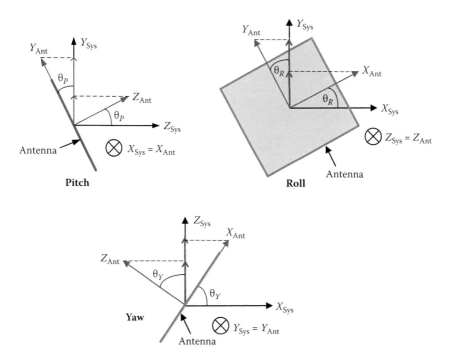

Figure 2.30 Roll, pitch, and yaw definitions.

For each of the different rotations, a transformation is required from the coordinate system of the array to that of the system. These transformations can be described using rotational matrices. The equations in (2.22) show the transformation matrices for roll, pitch, and yaw.

$$Roll = [R] = \begin{vmatrix} \cos(\theta_R) & -\sin(\theta_R) & 0 \\ \sin(\theta_R) & \cos(\theta_R) & 0 \\ 0 & 0 & 1 \end{vmatrix}$$

$$Pitch = [P] = \begin{vmatrix} 1 & 0 & 0 \\ 0 & \cos(\theta_P) & \sin(\theta_P) \\ 0 & -\sin(\theta_P) & \cos(\theta_P) \end{vmatrix} \qquad (2.22)$$

$$Yaw = [Y] = \begin{vmatrix} \cos(\theta_Y) & 0 & -\sin(\theta_Y) \\ 0 & 1 & 0 \\ \sin(\theta_Y) & 0 & \cos(\theta_Y) \end{vmatrix}$$

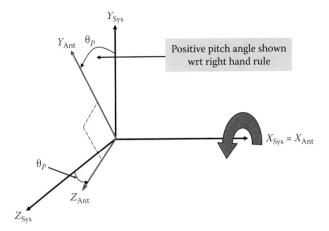

Figure 2.31 Right-hand rule dictates the direction of the roll, pitch, and yaw rotations.

These rotation matrices assume the roll, pitch, and yaw angle directions are based upon the right-hand rule. As an example, for a yaw rotation, the cross product of the unit vectors in the z and x axes, respectively, generates a y directed unit vector that is positive. This is illustrated in Figure 2.31.

Figure 2.32 shows a plot of a pattern without any pitch and a plot of a pattern with 30° pitch. As expected, the main beam of the pattern is shifted up 30° due to the pitch of the ESA. Corresponding plots for arbitrary roll, pitch, and yaw can be generated using the matrices defined in Equation (2.22). It is important to note that the order of the roll, pitch, and yaw matters. For example, $[R] \cdot [P] \neq [P] \times [R]$.

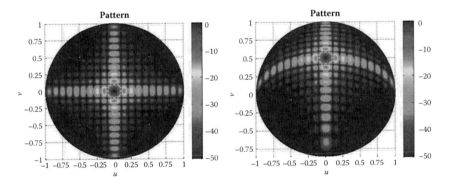

Figure 2.32 Effects of pitch on an ESA pattern. See insert.

2.5.3   Integrated Gain

The directive gain of an antenna is the ratio of the radiation intensity (power per unit solid angle) in a particular direction to the average power radiated over all space (Balanis 1982). The equation for the directive gain is

$$D(\theta,\phi) = \frac{4\pi\; U(\theta,\phi)}{\int_{\phi=0}^{2\pi}\int_{\theta=0}^{\pi} U(\theta,\phi)\sin\theta\; d\theta\; d\phi} \tag{2.23}$$

The directivity of an antenna is the maximum directive gain value and can be written as

$$D = \max(D(\theta,\phi)) \tag{2.24}$$

An alternative and commonly used expression for the directivity is

$$D = \frac{4\pi A}{\lambda^2}\cdot TL \tag{2.25}$$

For large arrays, Equation (2.25) can be used. However, for small arrays Equation (2.24) provides a more accurate result. Figure 2.33 illustrates the integrated gain for the pattern in Figure 2.19. Using Equation (2.25) the peak directivity is 31 dB. The simulated integrated gain value is 31.04 dB.

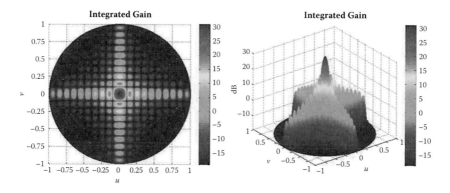

Figure 2.33 Integrated gain of the pattern in Figure 2.20 boresite antenna pattern (no electronic scan) in sine space. See insert.

2.6   *MATLAB Program and Function Listings*

This section contains a listing of all MATLAB® programs and functions used in this chapter.

2.6.1   *Compute_2D_AF.m (Function)*

```
%% Function to Compute 2D AF
% Arik D. Brown

function [AF, AF_mag, AF_dB, AF_dBnorm] =...
 Compute_2D_AF(wgts,nelemsx,nelemsy,dx_in,dy_in,f_GHz,fo_
 GHz,...
 randerror_amp,randerror_phs,u,v,uo,vo)

lambda=11.803/f_GHz;%wavelength(in)
lambdao=11.803/fo_GHz;%wavelength at tune freq(in)

k=2*pi/lambda;%rad/in
ko=2*pi/lambdao;%rad/in

AF=zeros(size(u));

phasex_mat=k*u-ko*uo;
phasey_mat=k*v-ko*vo;
randerror=randerror_amp.*exp(1j*randerror_phs*pi()/180);
for ii=1:nelemsx
 dx=(ii-(nelemsx+1)/2)*dx_in;
 for jj=1:nelemsy
 dy=(jj-(nelemsy+1)/2)*dy_in;
 AF = AF + wgts(ii,jj).*randerror(jj,ii)*...
  exp(1j*dx*phasex_mat).*exp(1j*dy*phasey_mat);
 end
end
[AF_mag AF_dB AF_dBnorm] = process_matrix(AF);
```

2.6.2   *Compute_2D_AFquant.m (Function)*

```
%% Function to Compute 2D AF with Quantization
% Arik D. Brown

function [AF, AF_mag, AF_dB, AF_dBnorm] =...
 Compute_2D_AF_quant(wgts,nelemsx,nelemsy,dx_in,dy_in,f_
 GHz,fo_GHz,...
 randerror_amp,randerror_phs,nbits,u,v,uo,vo)

lambda=11.803/f_GHz;%wavelength(in)
lambdao=11.803/fo_GHz;%wavelength at tune freq(in)
```

```
k=2*pi/lambda;%rad/in
ko=2*pi/lambdao;%rad/in

AF=zeros(size(u));

xpos_vec=([1:nelemsy]-(nelemsx+1)/2)*dx_in;
ypos_vec=([1:nelemsy]-(nelemsy+1)/2)*dy_in;
[xpos_mat,ypos_mat]=meshgrid(xpos_vec,ypos_vec);

LSB=360/(2^nbits);

randerror=randerror_amp.*exp(1j*randerror_phs*pi()/180);
for ii=1:nelemsx
 for jj=1:nelemsy
  phase1=k*(xpos_mat(ii,jj)*u+ypos_mat(ii,jj)*v);
  phase2=-ko*(180/pi)*(xpos_mat(ii,jj)*uo+ypos_
  mat(ii,jj)*vo);
  phase2_quant1=phase2/LSB;
  quant_delta=phase2_quant1-floor(phase2_quant1);
  if quant_delta <= 0.5
   phase2_quant2=floor(phase2_quant1)*LSB;
  elseif quant_delta > 0.5
   phase2_quant2=ceil(phase2_quant1)*LSB;
  end

 AF = AF + wgts(ii,jj).*randerror(jj,ii)*...
  exp(1j*phase1).*exp(1j*phase2_quant2*pi/180);
 end
end

[AF_mag AF_dB AF_dBnorm] = process_matrix(AF);
```

2.6.3   Compute_2D_EP.m (Function)

```
%% Function to Compute 1D EP
% Arik D. Brown

function [EP, EP_mag, EP_dB, EP_dBnorm] =...
 Compute_2D_EP(theta_deg,EF)

EP=zeros(size(theta_deg));

EP=(cosd(theta_deg).^(EF/2));%Volts
[EP_mag, EP_dB, EP_dBnorm] = process_matrix(EP);
```

2.6.4   Compute_2D_PAT.m (Function)

```
%% Function to Compute 2D PAT
% Arik D. Brown
```

```
function [PAT, PAT_mag, PAT_dB, PAT_dBnorm] =...
 Compute_2D_PAT(EP,AF)

PAT=zeros(size(AF));

PAT=EP.*AF;
[PAT_mag PAT_dB PAT_dBnorm] =...
 process_matrix(PAT);
```

2.6.5   Compute_2D_INTGAIN.m (Function)

```
%% Function to Compute Integrated Gain
% Arik D. Brown

function [PeakGain IdealGain PeakGaindB IdealGaindB
GainPattern_mag...
 GainPattern_dB GainPattern_dBnorm] =...
 Compute_2D_INTGAIN(Pattern_mag,thetavec,phivec,thetamat,...
 nelemx,nelemy,dx_in,dy_in,fGHz)

%% Compute Integrated Gain
numptstheta=length(thetavec);
numptsphi=length(phivec);
thetarad=thetavec*pi/180;
phirad=phivec*pi/180;
thetamat_rad=thetamat*pi/180;

dphi=(phirad(length(phirad))-phirad(1))/numptsphi;
dtheta=(thetarad(length(phirad))-thetarad(1))/numptstheta;
dsintheta=abs(sin(thetamat_rad));

GainPattern=(sum(sum((dsintheta*dphi*dtheta))))*2*Patt
ern_mag.^2/...
 (sum(sum( (Pattern_mag.^2) .*(dsintheta*dphi*dtheta) )));
PeakGain=max(max(GainPattern));
PeakGaindB=10*log10(PeakGain);
[GainPattern_mag GainPattern_dB GainPattern_dBnorm] =...
 process_matrix2(GainPattern);

%% Compute Ideal Gain
Area=nelemx*nelemy*dx_in*dy_in;
lambda=11.803/fGHz;
IdealGain=(4*pi*Area)/lambda.^2;
IdealGaindB=10*log10(IdealGain);
```

2.6.6   process_matrix.m (Function)

```
function[matrixmag,matrixdB,matrixdBnorm] = process_
matrix(matrix)
```

```
matrixmag=abs(matrix);
matrixmax=max(max(matrixmag));
matrixdB=20*log10(matrixmag+eps);
matrixdBnorm=matrixdB - 20*log10(matrixmax);
```

2.6.7   process_matrix2.m (Function)

```
function[matrixmag,matrixdB,matrixdBnorm] = process_
matrix2(matrix)

matrixmag=abs(matrix);
matrixmax=max(max(matrixmag));
matrixdB=10*log10(matrixmag+eps);
matrixdBnorm=matrixdB - 10*log10(matrixmax);
```

2.6.8   Taylor.m (Function)

```
function [wgt] = Taylor(points,sll,nbar)

r    = 10^(abs(sll)/20);
a    = log(r+(r*r-1)^0.5) / pi;
sigma2 = nbar^2/(a*a+(nbar-0.5)^2);

%--Compute Fm, the Fourier coefficients of the weight set
for m=1:(nbar-1)
 for n=1:(nbar-1)
  f(n,1)=1-m*m/sigma2/(a*a+(n-0.5)*(n-0.5));
  if n ~= m
   f(n,2)=1/(1-m*m/n/n);
  end
  if n==m
   f(n,2)=1;
  end
 end
 g(1,1)=f(1,1);
 g(1,2)=f(1,2);

 for n=2:(nbar-1)
  g(n,1)=g(n-1,1)*f(n,1);
  g(n,2)=g(n-1,2)*f(n,2);
 end
 F(m)=((-1)^(m+1))/2*g(nbar-1,1)*g(nbar-1,2);
end

jj = [1:points]';
xx = (jj-1+0.5)/points - 1/2; %-- column vector
W = ones(size(jj));   %-- column vector
mm = [1:nbar-1];    %-- row vector
```

```
W = W + 2*cos(2*pi*xx*mm)*F';

WPK = 1 + 2*sum(F);
wgt = W / WPK;
```

2.6.9   Pattern2D.m

```
% 2D Pattern Code
% Computes Element Pattern (EP), Array Factor(AF)and array
pattern (EP*AF)
% Arik D. Brown

clear all

%% Input Parameters
%ESA Parameters

%ESA opearating at tune freq
array_params.f=3;%Operating Frequency in GHz
array_params.fo=3;%Tune Frequency in GHz of the Phase Shifter,

array_params.nelem.x=20;%Number of Elements in x
array_params.nelem.y=20;%Number of Elements in y
array_params.d.x=(1/(1+sind(90)))*(11.803/array_params.
fo);%Element Spacing in Inches
array_params.d.y=(1/(1+sind(90)))*(11.803/array_params.
fo);%Element Spacing in Inches

array_params.EF=1.5;%EF

array_params.flag.gain=0;%0 = Don't Compute Gain, 1=Compute
Gain
array_params.flag.wgt=0;%0 = Uniform, 1 = Taylor Weighting
array_params.flag.error_rand=0;%0 = Ideal (no errors), 1 =
Random Phase/Amp. errors
array_params.flag.error_quant=0;%0 = Ideal (no quant.), 1
= Quant.
array_params.tilt.option=0;%0 = Ideal position (No roll,
pitch or yaw)
%1 = Roll (rotation about z axis)
%2 = Pitch (rotation about x axis)
%3 = Yaw (rotation about y axis)
array_params.tilt.angle=30;%degrees

array_params.error.rand.amp.onesigma=0.5;%dB
array_params.error.rand.phs.onesigma=6;%degrees
array_params.bits=6;
```

```
%$$$$These Parameters Only Used if array_params.wgtflag=1;
array_params.taylor.nbar=5;
array_params.taylor.SLL=30;%dB value

%Theta and Phi Angle Parameters (Antenna Coordinates)
theta_angle.numpts=361;%Number of angle pts in theta
phi_angle.numpts=361;%Number of angle pts in phi

theta_angle.min=0;%degrees
theta_angle.max=90;%degrees

phi_angle.min=0;%degrees
phi_angle.max=360;%degrees

theta_angle.scan=45;%degrees
phi_angle.scan=90;%degrees

plotcommand.coord=0;%0 = Antenna Coord., 1 = Radar
Coordinates
plotcommand.error=0;%0 = Don't Plot Errors, 1 = Plot Errors

plotcommand.EP=0;%Plot EP if = 1
plotcommand.AF=0;%Plot AF if = 1
plotcommand.PAT=1;%Plot PAT if = 1
plotcommand.INTGAIN=0;%Plot INTGAIN if = 1

array_params.dBfloor.EP=-20;% dB value for plotting
array_params.dBfloor.PAT=-50;% dB value for plotting

%% Computations

if array_params.flag.wgt==0
 array_params.amp_wgts.mat=ones(array_params.nelem.y,array_
 params.nelem.x);
else
 array_params.amp_wgts.vec.x=Taylor(array_params.
 nelem.x,array_params.taylor.SLL,...
 array_params.taylor.nbar);
 array_params.amp_wgts.vec.y=Taylor(array_params.
 nelem.y,array_params.taylor.SLL,...
 array_params.taylor.nbar);
 array_params.amp_wgts.mat=array_params.amp_wgts.vec.y*...
 array_params.amp_wgts.vec.x';
end

if array_params.flag.error_rand==0

 array_params.error.rand.amp.mat=ones(size(array_params.
 amp_wgts.mat));
```

```
array_params.error.rand.amp.mat_dB=10*log10(array_params.
error.rand.amp.mat);%dB

array_params.error.rand.phs.mat=zeros(size(array_params.
amp_wgts.mat));%degrees

elseif array_params.flag.error_rand==1

array_params.error.rand.amp.mat_dB=...
array_params.error.rand.amp.onesigma*randn(size(array_
params.amp_wgts.mat));%dB
array_params.error.rand.amp.mat=...
10.^(array_params.error.rand.amp.mat_dB/10);

array_params.error.rand.phs.mat=...
array_params.error.rand.phs.onesigma*randn(size(array_
params.amp_wgts.mat));%degrees

end

theta_angle.vec=linspace(theta_angle.min,theta_angle.max,...
 theta_angle.numpts);%degrees
phi_angle.vec=linspace(phi_angle.min,phi_angle.max,...
 phi_angle.numpts);%degrees

[theta_angle.mat phi_angle.mat]=meshgrid(theta_angle.
vec,phi_angle.vec);

if array_params.tilt.option == 0
 array_params.tilt_mat=[1 0 0;...
  0 1 0;...
  0 0 1];
elseif array_params.tilt.option == 1
 array_params.tilt_mat=[cosd(array_params.tilt.angle)
 -sind(array_params.tilt.angle) 0;...
  sind(array_params.tilt.angle) cosd(array_params.tilt.
  angle) 0;...
  0 0 1];
elseif array_params.tilt.option == 2
 array_params.tilt_mat=[1 0 0;...
  0 cosd(array_params.tilt.angle) sind(array_params.tilt.
  angle);...
  0 -sind(array_params.tilt.angle) cosd(array_params.tilt.
angle)];
elseif array_params.tilt.option == 3
 array_params.tilt_mat=[cosd(array_params.tilt.angle) 0
 -sind(array_params.tilt.angle);...
  0 1 0;...
```

```
   sind(array_params.tilt.angle) 0 cosd(array_params.tilt.
   angle)];
end

sinespace.umat=sind(theta_angle.mat).*cosd(phi_angle.mat);
sinespace.vmat=sind(theta_angle.mat).*sind(phi_angle.mat);
sinespace.wmat=sqrt(abs(1-(sinespace.umat.^2)-(sinespace.
vmat.^2)));

sinespace.uo=sind(theta_angle.scan)*cosd(phi_angle.scan);
sinespace.vo=sind(theta_angle.scan)*sind(phi_angle.scan);
sinespace.wo=sqrt(1-(sinespace.uo.^2)-(sinespace.vo.^2));

sinespace.uvwmat=[sinespace.umat(:).';sinespace.
vmat(:).';...
 sinespace.wmat(:).'];
sinespace.uvwmatnew=array_params.tilt_mat*sinespace.uvwmat;
sinespace.umatnew=reshape(sinespace.
uvwmatnew(1,:),size(sinespace.umat));
sinespace.vmatnew=reshape(sinespace.
uvwmatnew(2,:),size(sinespace.vmat));
sinespace.wmatnew=reshape(sinespace.
uvwmatnew(3,:),size(sinespace.wmat));

if plotcommand.coord==0
 plotxy.x=sinespace.umatnew;
 plotxy.y=sinespace.vmatnew;
 plotxy.xtxt='u';
 plotxy.ytxt='v';
 plotxy.xtick.min=-1;
 plotxy.xtick.max=1;
 plotxy.ytick.min=-1;
 plotxy.ytick.max=1;
 plotxy.tick.delta=.25;
elseif plotcommand.coord==1
 radarcoord.thetaAZmat=atan2(sinespace.umat,sinespace.
 wmat)*180/pi;%degrees
 radarcoord.thetaELmat=asind(sinespace.vmat);%degrees
 plotxy.x=radarcoord.thetaAZmat;
 plotxy.y=radarcoord.thetaELmat;
 plotxy.xtxt='\theta_{AZ}';
 plotxy.ytxt='\theta_{EL}';
 plotxy.xtick.min=-90;
 plotxy.xtick.max=90;
 plotxy.ytick.min=-90;
 plotxy.ytick.max=90;
 plotxy.tick.delta=30;
end
```

```
%Initialize Element Pattern, Array Factor and Pattern
array.size=size(theta_angle.mat);
array.EP=zeros(array.size);%EP
array.AF=zeros(array.size);%AF
array.PAT=zeros(array.size);
%% Compute Patterns

%Compute AF1

if array_params.flag.error_quant == 0
 [array.AF, array.AF_mag, array.AF_dB, array.AF_dBnorm]=...
  Compute_2D_AF(array_params.amp_wgts.mat,...
  array_params.nelem.x,array_params.nelem.y,...
  array_params.d.x,array_params.d.y,...
  array_params.f,array_params.fo,...
  array_params.error.rand.amp.mat,array_params.error.rand.
  phs.mat,...
  sinespace.umat,sinespace.vmat,...
  sinespace.uo,sinespace.vo);
elseif array_params.flag.error_quant == 1
 [array.AF, array.AF_mag, array.AF_dB, array.AF_dBnorm]=...
  Compute_2D_AF_quant(array_params.amp_wgts.mat,...
  array_params.nelem.x,array_params.nelem.y,...
  array_params.d.x,array_params.d.y,...
  array_params.f,array_params.fo,...
  array_params.error.rand.amp.mat,array_params.error.rand.
  phs.mat,...
  array_params.bits,...
  sinespace.umat,sinespace.vmat,...
  sinespace.uo,sinespace.vo);
end

%Compute EP
[array.EP, array.EP_mag, array.EP_dB, array.EP_dBnorm]=...
 Compute_2D_EP(theta_angle.mat,array_params.EF);

%Compute PAT
[array.PAT, array.PAT_mag, array.PAT_dB, array.PAT_dBnorm]
=...
 Compute_2D_PAT(array.EP,array.AF);

if array_params.flag.gain==1
 [array.INTGAINpeak array.IdealGain array.INTGAINpeakdB
 array.IdealGaindB...
  array.INTGAIN array.INTGAIN_dB array.INTGAIN_dBnorm] =...
  Compute_2D_INTGAIN(array.PAT_mag,...
  theta_angle.vec,theta_angle.vec,theta_angle.mat,...
  array_params.nelem.x,array_params.nelem.y,...
```

```
  array_params.d.x,array_params.d.y,...
  array_params.f);
 [array.INTGAINpeakdB array.IdealGaindB]
end

%% Plotting

% close all

if plotcommand.error == 1

 h=figure;clf
 set(gcf,'DefaultLineLineWidth',1.5)
 imagesc(array_params.error.rand.amp.mat_dB)
 shading interp
 set(gca,'FontSize',14,'FontWeight','bold')
 title('Random Amplitude Error(dB)')
 xlabel('Elements in x'),ylabel('Elements in y')
 view(2)
 colorbar
 caxis([-2 2])
 set(gca,'FontSize',14,'FontWeight','bold')
 set(gcf, 'color', 'white');

 h=figure;clf
 binvector=[-2:.1:2];
 set(gcf,'DefaultLineLineWidth',1.5)
 hist(array_params.error.rand.amp.mat_dB(:),binvector);
 set(gca,'FontSize',14,'FontWeight','bold')
 title('Amplitude Error Distribution')
 xlabel('Amplitude Error Bins (dB)')
 axis tight
 set(gca,'XTick',[-2:0.5:2])
 set(gcf, 'color', 'white');

 h=figure;clf
 set(gcf,'DefaultLineLineWidth',1.5)
 imagesc(array_params.error.rand.phs.mat)
 shading interp
 set(gca,'FontSize',14,'FontWeight','bold')
 title('Random Phase Error(^{o})')
 xlabel('Elements in x'),ylabel('Elements in y')
 view(2)
 colorbar
 caxis([-20 20])
 set(gca,'FontSize',14,'FontWeight','bold')
 set(gcf, 'color', 'white');
```

```
 h=figure;clf
 binvector=[-20:1:20];
 set(gcf,'DefaultLineLineWidth',1.5)
 hist(array_params.error.rand.phs.mat(:),binvector);
 set(gca,'FontSize',14,'FontWeight','bold')
 title('Phase Error Distribution')
 xlabel('Phase Error Bins (^{o})')
 axis tight
 set(gca,'XTick',[-20:5:20])
 set(gcf, 'color', 'white');

end

if plotcommand.EP == 1

 %Plot EP in dB, Normalized
 plotEP=array.EP_dBnorm;
 plotEP(array.EP_dBnorm < array_params.dBfloor.PAT)=array_
 params.dBfloor.PAT;

 h=figure;clf
 set(gcf,'DefaultLineLineWidth',1.5)
 surf(plotxy.x,plotxy.y,array.EP_dBnorm),hold
 shading interp
 colorbar
 caxis([array_params.dBfloor.EP 0])
 set(gca,'FontSize',14,'FontWeight','bold')
 title('Element Pattern')
 xlabel(plotxy.xtxt),ylabel(plotxy.ytxt),zlabel('dB')
 view(2)%Plot EP in dB, Normalized
 axis tight
 set(gca,'XTick',[plotxy.xtick.min:plotxy.tick.delta:plotxy.
 xtick.max])
 set(gca,'YTick',[plotxy.ytick.min:plotxy.tick.delta:plotxy.
 ytick.max])
 set(gcf, 'color', 'white');

 h=figure;clf
 set(gcf,'DefaultLineLineWidth',1.5)
 surf(plotxy.x,plotxy.y,plotEP),hold
 shading interp
 colorbar
 caxis([array_params.dBfloor.PAT 0])
 zlim([array_params.dBfloor.PAT 0])
 set(gca,'FontSize',14,'FontWeight','bold')
 title('Element Pattern')
 xlabel(plotxy.xtxt),ylabel(plotxy.ytxt),zlabel('dB')
 view(3)
```

```
set(gca,'XTick',[plotxy.xtick.min:plotxy.tick.delta:plotxy.
xtick.max])
set(gca,'YTick',[plotxy.ytick.min:plotxy.tick.delta:plotxy.
ytick.max])
set(gcf, 'color', 'white');
end

if plotcommand.AF == 1
 %Plot PAT in dB, Normalized
 plotAF=array.AF_dBnorm;
 plotAF(array.AF_dBnorm < array_params.dBfloor.PAT)=array_
 params.dBfloor.PAT;

 h=figure;clf
 set(gcf,'DefaultLineLineWidth',1.5)
 surf(plotxy.x,plotxy.y,array.AF_dBnorm)
 shading interp
 colorbar
 caxis([array_params.dBfloor.PAT 0])
 set(gca,'FontSize',14,'FontWeight','bold')
 title('AF')
 xlabel(plotxy.xtxt),ylabel(plotxy.ytxt),zlabel('dB')
 view(2)
 axis tight
 set(gca,'XTick',[plotxy.xtick.min:plotxy.tick.delta:plotxy.
 xtick.max])
 set(gca,'YTick',[plotxy.ytick.min:plotxy.tick.delta:plotxy.
 ytick.max])
 set(gcf, 'color', 'white');

 %Plot PAT in dB, Normalized
 h=figure;clf
 set(gcf,'DefaultLineLineWidth',1.5)
 surf(plotxy.x,plotxy.y,plotAF)
 shading interp
 colorbar
 caxis([array_params.dBfloor.PAT 0])
 zlim([array_params.dBfloor.PAT 0])
 set(gca,'FontSize',14,'FontWeight','bold')
 title('AF')
 xlabel(plotxy.xtxt),ylabel(plotxy.ytxt),zlabel('dB')
 view(3)
 set(gca,'XTick',[plotxy.xtick.min:plotxy.tick.delta:plotxy.
 xtick.max])
 set(gca,'YTick',[plotxy.ytick.min:plotxy.tick.delta:plotxy.
 ytick.max])
 set(gcf, 'color', 'white');
end
```

```
if plotcommand.PAT == 1
 %Plot PAT in dB, Normalized
 plotPAT=array.PAT_dBnorm;
 plotPAT(array.PAT_dBnorm <= array_params.dBfloor.
 PAT)=array_params.dBfloor.PAT;

 h=figure;clf
 set(gcf,'DefaultLineLineWidth',1.5)
 surf(plotxy.x,plotxy.y,array.PAT_dBnorm)
 shading interp
 colorbar
 caxis([array_params.dBfloor.PAT 0])
 set(gca,'FontSize',14,'FontWeight','bold')
 title('Pattern')
 xlabel(plotxy.xtxt),ylabel(plotxy.ytxt),zlabel('dB')
 view(2)
 set(gca,'XTick',[plotxy.xtick.min:plotxy.tick.delta:plotxy.
 xtick.max])
 set(gca,'YTick',[plotxy.ytick.min:plotxy.tick.delta:plotxy.
 ytick.m axis tight
 set(gcf, 'color', 'white');

 %Plot PAT in dB, Normalized
 h=figure;clf
 set(gcf,'DefaultLineLineWidth',1.5)
 surf(plotxy.x,plotxy.y,plotPAT)
 shading interp
 colorbar
 caxis([array_params.dBfloor.PAT 0])
 zlim([array_params.dBfloor.PAT 0])
 set(gca,'FontSize',14,'FontWeight','bold')
 title('Pattern')
 xlabel(plotxy.xtxt),ylabel(plotxy.ytxt),zlabel('dB')
 view(3)
 set(gca,'XTick',[plotxy.xtick.min:plotxy.tick.delta:plotxy.
 xtick.max])
 set(gca,'YTick',[plotxy.ytick.min:plotxy.tick.delta:plotxy.
 ytick.max])
 set(gcf, 'color', 'white');
end

if array_params.flag.gain == 1 && plotcommand.INTGAIN == 1
 %Plot INTGAIN in dB, Normalized
 plotINTGAIN=array.INTGAIN_dB;
 plotINTGAIN(array.INTGAIN_dB <= array.INTGAINpeakdB+array_
 params.dBfloor.PAT)=...
 array.INTGAINpeakdB+array_params.dBfloor.PAT;
```

```
h=figure;clf
set(gcf,'DefaultLineLineWidth',1.5)
surf(plotxy.x,plotxy.y,array.INTGAIN_dB)
shading interp
colorbar
caxis([array.INTGAINpeakdB+array_params.dBfloor.PAT array.
INTGAINpeakdB])
set(gca,'FontSize',14,'FontWeight','bold')
title('Integrated Gain')
xlabel(plotxy.xtxt),ylabel(plotxy.ytxt),zlabel('dB')
view(2)
set(gca,'XTick',[plotxy.xtick.min:plotxy.tick.delta:plotxy.
xtick.max])
set(gca,'YTick',[plotxy.ytick.min:plotxy.tick.delta:plotxy.
ytick.max])
axis tight
set(gcf, 'color', 'white');

%Plot INTGAIN in dB, Normalized
h=figure;clf
set(gcf,'DefaultLineLineWidth',1.5)
surf(plotxy.x,plotxy.y,plotINTGAIN)
shading interp
colorbar
caxis([array.INTGAINpeakdB+array_params.dBfloor.PAT array.
INTGAINpeakdB])
zlim([array.INTGAINpeakdB+array_params.dBfloor.PAT array.
INTGAINpeakdB])
set(gca,'FontSize',14,'FontWeight','bold')
title('Integrated Gain')
xlabel(plotxy.xtxt),ylabel(plotxy.ytxt),zlabel('dB')
view(3)
set(gca,'XTick',[plotxy.xtick.min:plotxy.tick.delta:plotxy.
xtick.max])
set(gca,'YTick',[plotxy.ytick.min:plotxy.tick.delta:plotxy.
ytick.max])
set(gcf, 'color', 'white');
end
```

2.6.10   *GratingLobePlotter.m*

```
%% Code to compute and plot grating lobes in sine space
%Arik D. Brown

%% Define input parameters
c=11.803;%Gin/s
f=3;%GHz
lambda=c/f;%in
```

```
%Element Spacing
dx=.5*lambda;%in
dy=.29*lambda;%in

%Grid Type
elemgrid=2;%1 = Rectangular Grid, 2 = Triangular Grid

%Define Scan Angle
thetaELo=0;%deg thetaEL
thetaAZo=0;%deg phiAZ
mainbeam.uo=cosd(thetaELo)*sind(thetaAZo);
mainbeam.vo=sind(thetaELo);

%Define Unit Circle
phi=[0:1:360];%degrees
unitcircle.u=cosd(phi);
unitcircle.v=sind(phi);

%Define Main Beam Unit Circle
mainbeam.u=mainbeam.uo + unitcircle.u;
mainbeam.v=mainbeam.vo + unitcircle.v;

%Define Grating Lobe Locations
if elemgrid == 1
 gratinglobe.uo=mainbeam.uo+[ 0 -1 -1 1 1 0 -1 1]*lambda/dx;
 gratinglobe.vo=mainbeam.vo+[ 1 -1 0 0 1 -1 1 -1]*lambda/dy;
 plotfigs.titletxt='Rectangular Grid';
 plotfigs.markertype='s';
elseif elemgrid ==2
 gratinglobe.uo=mainbeam.uo+[ 0 0 1 -1 -1 1 2 -2]*lambda/
 (2*dx);
 gratinglobe.vo=mainbeam.vo+[ 2 -2 1 -1 1 -1 0 0]*lambda/
 (2*dy);
 plotfigs.titletxt='Triangular Grid';
 plotfigs.markertype='^';
end

%% Plot Grating Lobes

plotfigs.limit=4.5;

figure 1)
clf

%Mainbeam Unit Circle
plot(mainbeam.u,mainbeam.v,'-','color',[0 0
0],'LineWidth',1.5),hold
```

```
%Main Beam
plot(mainbeam.uo,mainbeam.vo,plotfigs.markertype,'color',[0
0 0],'MarkerSize',6,'LineWidth',1.5)

%Grating Lobes
plot(gratinglobe.uo,gratinglobe.vo,plotfigs.
markertype,'color',[1 0 0],'LineWidth',1.5)
%Grating Lobe Unit Circles
for ig=1:length(gratinglobe.uo)

 if elemgrid == 1
  plot(unitcircle.u+ gratinglobe.uo(ig),unitcircle.
  v+gratinglobe.vo(ig),'r-','LineWidth',1.5)
 elseif elemgrid == 2
  plot(unitcircle.u+ gratinglobe.uo(ig),unitcircle.
  v+gratinglobe.vo(ig),'r-','LineWidth',1.5)
 end
end

axis([-plotfigs.limit plotfigs.limit -plotfigs.limit plot-
figs.limit])
set(gca,'XTick',[-plotfigs.limit:.5:plotfigs.
limit],'YTick',[-plotfigs.limit:.5:plotfigs.limit])
xlabel('u','FontSize',14,'FontWeight','bold')
ylabel('v','FontSize',14,'FontWeight','bold')
title(plotfigs.titletxt,'FontSize',18,'FontWeight','bold')
grid
set(gca,'Fontsize',14,'Fontweight','bold','linewidth',1.0)
```

References

Balanis, Constantine. *Antenna Theory: Analysis and Design.* New York: John Wiley & Sons, Publishers, 1982.

Batzel, Urs, Kurt Ramsey, and Daniel Davis. *Angular Coordinate Definitions.* Antenna Fundamentals Class. Baltimore, MD: Northrop Grumman Electronic Systems, August 2010.

Davis, Daniel. *Grating Lobes Analysis.* Antenna Fundamentals Class. Baltimore, MD: Northrop Grumman Electronic Systems, July 2010.

Idstein, Kevin. *Coordinate Transforms.* Antenna Fundamentals Class. Baltimore, MD: Northrop Grumman Electronic Systems, 2010.

Konapelsky, Richard. *Coordinate Transformations.* Antenna Systems Class. Baltimore, MD: Northrop Grumman Electronic Systems, 2002.

Long, John, and Kevin Schmidt. *Presenting Antenna Patterns with FFT and DFT.* Antenna Fundamentals Class. Baltimore, MD: Northrop Grumman Electronic Systems, July 2010.

Mailloux, Robert J. *Phased Array Antenna Handbook.* Norwood, MA: Artech House, 1994.

Miller, C. J. Minimizing the Effects of Phase Quantization Errors in an Electronically Scanned Array. In *Proceedings of 1964 Symposium on Electronically Scanned Phased Arrays and Applications*, 1964, 17–38.

Skolnik, Merrill. *Radar Handbook*, 2nd ed. McGraw-Hill, 1990.

chapter three

Subarray Beamforming

Arik D. Brown
Northrop Grumman Electronic Systems

Contents

3.1 Introduction

In Chapters 1 and 2, the basic fundamentals of ESAs are discussed. Important topics germane to ESA design are covered, such as beamwidth, instantaneous bandwidth (IBW), grating lobes, etc. This chapter builds upon these fundamentals. In Chapter 1, the topic of IBW was covered in detail. It was shown in Section 1.3.2 that the IBW is inversely proportional to the length of an aperture. For most current and emerging systems that require ESAs, larger IBWs are desired on the order of hundreds of MHz. For a given aperture size, the IBW is fixed. If the IBW is less than that desired, then the ESA's performance will suffer from beam squint. This is depicted in Figure 3.1.

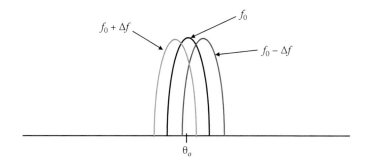

Figure 3.1 Undesired ESA beam squint due to an aperture length that does not support a desired IBW. (From Wojtowicz, J., *Instantaneous Bandwidth Analysis and Subarrayed Antennas*, Antenna Systems Class, Northrop Grumman Electronic Systems, Baltimore, MD, April 2000.)

Additionally, for applications where greater sensitivity is required via larger ESA apertures (on the order of 1 m² or larger), IBW is limited. Larger apertures will inherently have limited IBW.

The discussion in the preceding paragraph assumes phase shifter steering at each ESA element with an analog beamformer used to coherently add the signal from all the elements. This ESA architecture topology is illustrated in Figure 3.2. In order to provide, wide IBWs' true time delay could be used at each element in place of the phase shifters shown in Figure 3.2. However, time delay at each element would be expensive, lossy, and contain an additional source of error (Skolnik 1990). The practical solution is to use subarrays that are multiple single elements grouped

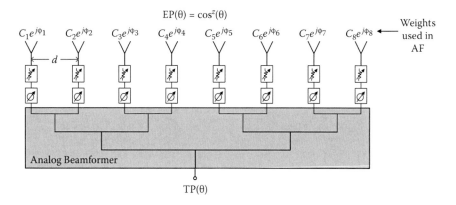

Figure 3.2 ESA topology employing phase shifters at each element and a passive manifold combiner. (From Ulrey, J., *Sub Arrays and Beamforming*, Antenna Fundamentals Class, Northrop Grumman Electronic Systems, Baltimore, MD, 2010.)

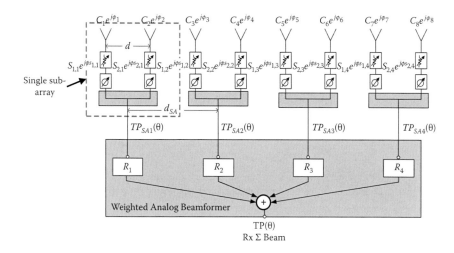

Figure 3.3 ESA subarray topology with the passive manifold combining done on a subarray level as opposed to element to element in Figure 3.2. (From Ulrey, J., *Sub Arrays and Beamforming*, Antenna Fundamentals Class, Northrop Grumman Electronic Systems, Baltimore, MD, 2010.)

together. The subarrays form the entire array and the subarrays can be thought of as having the effective element pattern that is used to compute the entire pattern. A subarray architecture topology is shown in Figure 3.3. The analog beamformer in Figure 3.3 performs the combining on a subarray-to-subarray level in contrast to the element-to-element combining illustrated in Figure 3.2.

In addition to alleviating the issues discussed previously in this section related to IBW, subarray architectures can also reduce the number of control elements required in an ESA. For applications that have limited scan requirements phase shifters can be employed behind groups of subarrays as opposed to each element. However this provides very limited scan performance, as will be discussed later in this chapter. Another benefit subarrays provide is a larger building block for manufacturing and fabrication. Instead of building many individual modules with a single phase shifter, subarray building blocks can be built, which is quite attractive for larger arrays.

3.2   Subarray Pattern Formulation

This chapter will focus on the one-dimensional (1D) formulation to explain and describe ESA pattern behavior for subarray ESA architectures. The two-dimensional (2D) formulation for subarrays is left as an exercise for the reader, but can be readily derived as outlined by the formulation shown in Chapter 2 for nonsubarrayed ESAs. The 1D formulation is more than satisfactory to provide the insight into subarray ESA pattern performance.

In Chapter 1, the pattern for a 1D ESA is shown to be

$$F(\theta) = \cos^{\frac{EF}{2}} \theta \cdot \sum_{m=1}^{M} a_m e^{j\left(\frac{2\pi}{\lambda}x_m \sin\theta - \frac{2\pi}{\lambda_o}x_m \sin\theta_o\right)} \tag{3.1}$$

For ease of notation the element pattern will be referred to as *EP* for the remainder of this chapter. For a subarrayed architecture, the pattern formulation is similar to Equation (3.1) and can be expressed in the same manner. We begin by assuming an ESA comprised of *M* individual elements. These *M* elements are divided into *P* subarrays where the number of elements per subarray, *R*, is $\frac{M}{P}$. As an example, the architecture pictured in Figure 3.3 has $M = 8$, $P = 4$, and $R = 2$. The number of elements in the array does not have to be an integer multiple of the subarray elements. In the case of overlapped subarrays, the subarrays on the edge of the array may have fewer elements per subarray. This will be shown later in this chapter.

Using Equation (3.1), the pattern for the subarrays can be written as

$$F_{SA}(\theta) = \cos^{\frac{EF}{2}} \theta \cdot \sum_{r=1}^{R} a_r e^{j\left(\frac{2\pi}{\lambda}x_r \sin\theta - \frac{2\pi}{\lambda_o}x_r \sin\theta_o\right)} \tag{3.2}$$

In order to express the complete pattern, a term needs to be added to Equation (3.2) for the analog beamforming. This results in the following expression:

$$F(\theta) = F_{SA}(\theta) \cdot \sum_{p=1}^{P} b_p e^{j\left(\frac{2\pi}{\lambda}x_p \sin\theta - \frac{2\pi}{\lambda_o}x_p \sin\theta_o\right)} \tag{3.3}$$

Combining Equations (3.2) and (3.3), and substituting AF_{SA} for the summation in Equation (3.2), the expression for a subarrayed ESA pattern is the following:

$$F(\theta) = EP \cdot \sum_{p=1}^{P} b_p \cdot AF_{SA_p} e^{j\left(\frac{2\pi}{\lambda}x_p \sin\theta - \frac{2\pi}{\lambda_o}x_p \sin\theta_o\right)} \tag{3.4}$$

If we assume that all the subarrays have the same AF, Equation (3.4) can be further simplified as

$$F(\theta) = (EP \cdot AF_{SA}) \cdot \sum_{p=1}^{P} b_p \cdot e^{j\left(\frac{2\pi}{\lambda}x_p \sin\theta - \frac{2\pi}{\lambda_o}x_p \sin\theta_o\right)} \tag{3.5}$$

Equation (3.5) closely resembles Equation (3.1). The difference is that the effective element pattern is the element pattern of a single element multiplied by the AF. Additionally, the summation is done over the number of subarrays instead of the number of elements. Simplifying Equation (3.5) more we arrive at the following expression:

$$F(\theta) = (EP \cdot AF_{SA}) \cdot AF_P = EP_{SA} \cdot AF_P \tag{3.6}$$

Equation (3.6) is a very intuitive result. It shows that the overall pattern for a subarrayed ESA reduces to pattern multiplication of the subarray element pattern with the AF of the backend manifolding. This is very useful in understanding the implications of using true time delay or digital beamforming across the subarrays, instead of using phase shifters as shown in Equation (3.5). These implications will be elaborated on in the following section.

3.3   Subarray Beamforming

There are multiple ESA architecture topologies that can be used for implementing subarray beamforming. These involve using one or multiple combinations of phase shifter, time delay, and digital beamforming. In this chapter, we will discuss three different architecture approaches with the limitations and advantage that each one provides. These architectures are illustrated in Figure 3.4 and will be expounded upon in the following subsections.

3.3.1   Subarray Phase Shifter Beamforming

Figure 3.5 shows an ESA subarray architecture employing only phase shifters at the subarray level. As mentioned earlier in this chapter, this is advantageous because the number of control elements required can be dramatically reduced. Instead of using a phase shifter at each element to electronically scan the beam, phase shifters are used only at the subarray level. Expanding Equation (3.5), the equation for a subarrayed pattern is

$$F(\theta) = \left(EP \cdot \sum_{r=1}^{R} a_r e^{j\left(\frac{2\pi}{\lambda} x_r \sin\theta - \frac{2\pi}{\lambda_o} x_r \sin\theta_o\right)} \right) \cdot \sum_{p=1}^{P} b_p \cdot e^{j\left(\frac{2\pi}{\lambda} x_p \sin\theta - \frac{2\pi}{\lambda_o} x_p \sin\theta_o\right)} \tag{3.7}$$

For the array in Figure 3.5, the array factor term in Equation (3.7) must be modified because there are no phase shifters on the element level. Equation (3.7) then becomes

$$F(\theta) = \left(EP \cdot \sum_{r=1}^{R} e^{j\left(\frac{2\pi}{\lambda} x_r \sin\theta\right)} \right) \cdot \sum_{p=1}^{P} b_p \cdot e^{j\left(\frac{2\pi}{\lambda} x_p \sin\theta - \frac{2\pi}{\lambda_o} x_p \sin\theta_o\right)} \tag{3.8}$$

Figure 3.6 illustrates the expression in Equation (3.7).

Limited Scan   **Limited IBW**

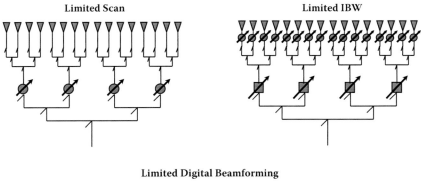

Limited Digital Beamforming

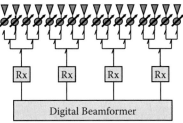

Figure 3.4 Three different ESA topologies that incorporate subarray beamforming. (From Wojtowicz, J., *Instantaneous Bandwidth Analysis and Subarrayed Antennas*, Antenna Systems Class, Northrop Grumman Electronic Systems, Baltimore, MD, April 2000.)

Limited Scan

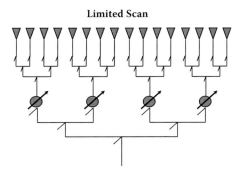

Figure 3.5 Electronic steering accomplished via phase delay implemented only at the subarray level. (From Wojtowicz, J., *Instantaneous Bandwidth Analysis and Subarrayed Antennas*, Antenna Systems Class, Northrop Grumman Electronic Systems, Baltimore, MD, April 2000.)

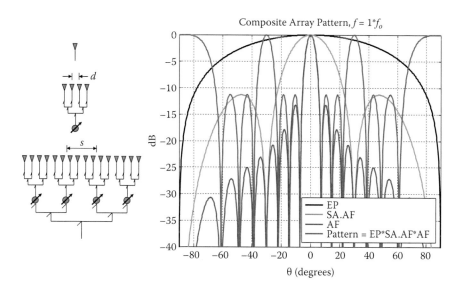

Figure 3.6 A pictorial representation of pattern multiplication for a subar-rayed ESA with phase-only steering at the subarray level. (From Wojtowicz, J., *Instantaneous Bandwidth Analysis and Subarrayed Antennas*, Antenna Systems Class, Northrop Grumman Electronic Systems, Baltimore, MD, April 2000.)

Although the number of control elements has been reduced the scan capability has also been reduced. Because the AF of the subarray does not steer with the beam, unwanted high sidelobes are the result in the pattern. This is shown in Figure 3.7. Phase shifter steering only at the sub-array level is then only viable for applications where the scan requirement is limited. It is important to point out that even with time delay at the subarray level instead of phase delay, the results are the same. With no elemental control, the AF of the subarray cannot be electronically steered and thus limits the ESA's ability to scan.

3.3.2 Subarray Time Delay Beamforming

The previous section showed that in order to not restrict the scan capa-bility of a subarrayed ESA, some type of elemental control is required. This allows the subarray AF to scan with the main beam. Phase shifter delay or time delay can be employed on the element level. However, elemental time delay adds undesired complexity especially for large ESAs (hundreds or thousands of elements). This leads to the approach

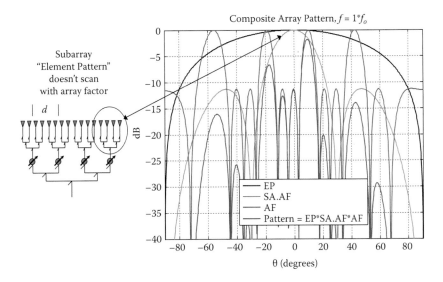

Figure 3.7 Without phase shifter control at the element level, the AF of the sub-array does not electronically steer with the beam and limits the scan capability. (From Wojtowicz, J., *Instantaneous Bandwidth Analysis and Subarrayed Antennas*, Antenna Systems Class, Northrop Grumman Electronic Systems, Baltimore, MD, April 2000.)

shown in Figure 3.8. The expression for the full array pattern using Equation (3.7) is

$$F(\theta) = \left(EP \cdot \sum_{r=1}^{R} a_r e^{j\left(\frac{2\pi}{\lambda} x_r \sin\theta - \frac{2\pi}{\lambda_o} x_r \sin\theta_o\right)} \right) \cdot \sum_{p=1}^{P} b_p \cdot e^{j\frac{2\pi}{\lambda} x_p (\sin\theta - \sin\theta_o)} \tag{3.9}$$

Using phase delay at the element level and time delay at the subarray level reduces the number of time delay devices required and provides excellent scan performance. Implementing control on the element level allows the AF of the subarray to be scanned with the array. Figure 3.9 shows an example of scanning the ESA with the approach shown in Figure 3.8. The array is scanned to 30° at the tune frequency, f_o, and the pattern is well behaved, as is expected.

A drawback of this approach is the limited IBW that results. Looking at Equation (3.9), the AF of the subarray (in parentheses) has a maximum value at the tune frequency (i.e., $\lambda = \lambda_o$); however, at frequencies offset from the tune frequency the AF squints and does not have a maximum value at the off-tune frequency (i.e., $\lambda \neq \lambda_o$). This is illustrated in Figure 3.10. This

Limited IBW

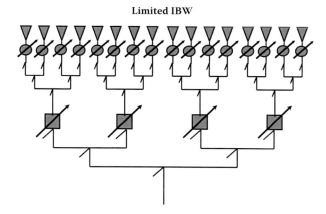

Figure 3.8 Electronic steering accomplished via phase delay at the element level and time delay at the subarray level. (From Wojtowicz, J., *Instantaneous Bandwidth Analysis and Subarrayed Antennas*, Antenna Systems Class, Northrop Grumman Electronic Systems, Baltimore, MD, April 2000.)

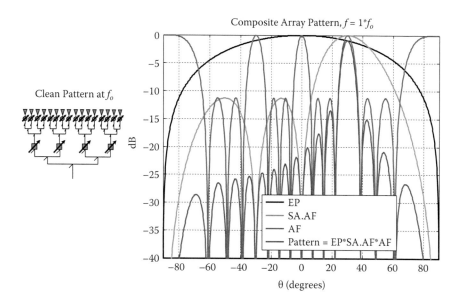

Figure 3.9 Using phase and time delay removes the scan limitations imposed by phase steering only at the subarray level. (Wojtowicz 2000.)

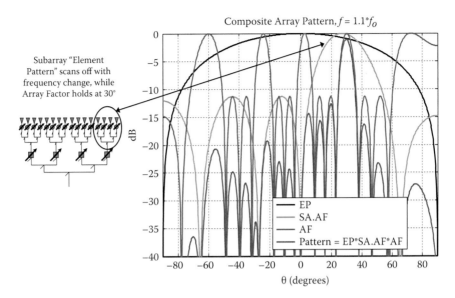

Figure 3.10 The architecture approach in Figure 3.8 has IBW limitations. (Wojtowicz 2000.)

limitation in IBW is still better than a nonsubarrayed ESA. In Chapter 1, the IBW of an ESA was shown to be

$$IBW = \frac{c}{L \sin \theta_o} \tag{3.10}$$

For a subarrayed ESA with time delay at the subarray level, the IBW is limited by the size of the subarray and not the array and can be expressed as

$$IBW = \frac{c}{L_{SA} \sin \theta_o}$$

$$= \frac{Pc}{L \sin \theta_o} \tag{3.11}$$

where P is the number of subarrays. Comparing Equations (3.10) and (3.11) the IBW is increased by the number of subarrays P. However, with the increased IBW there is an accompanying gain loss (no free lunch!) which is described in Skolnik (1990) as

$$Loss\ in\ gain \approx 1 - \left(\frac{\sin\left[\left(\pi/4 \right) \sin \theta_o \right]}{\left(\pi/4 \right) \sin \theta_o} \right)^2 \tag{3.12}$$

3.3.3   Subarray Digital Beamforming

Implementing phase delay at the element level and time delay at the subarray level provides robust scan performance with limitations on IBW. These limitations still provide an advantage over a nonsubarrayed ESA of the same size that has phase-only steering. An alternative to time delay steering at the subarray level is to place a receiver channel at each subarray and combine the subarrays digitally. This is referred to as digital beamforming and is shown in Figure 3.11. Mathematically, the pattern formulation is similar to that shown in Equation (3.9).

Digital beamforming (DBF) is an enabler for generating multiple simultaneous ESA beams with full aperture gain. By adjusting the digital weights, multiple beams can be created simultaneously. This is extremely beneficial because each of these beams has full aperture gain. This is illustrated in Figure 3.12. The same issues with IBW exist with DBF. In order to minimize this effect, the subarray aperture length is designed to match the IBW required.

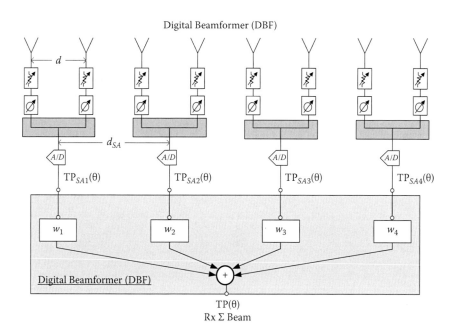

Figure 3.11 Digital beamforming at the subarray level. (Ulrey 2010.)

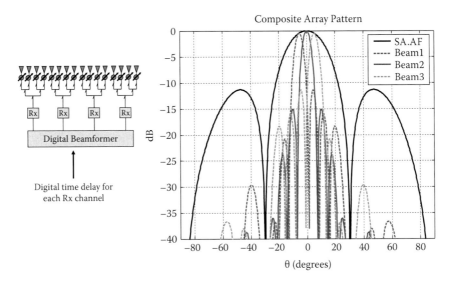

Figure 3.12 Digital beamforming allows multiple simultaneous beams with full aperture gain. (From Wojtowicz, J., *Instantaneous Bandwidth Analysis and Subarrayed Antennas*, Antenna Systems Class, Northrop Grumman Electronic Systems, Baltimore, MD, April 2000.)

3.4   Overlapped Subarrays

For both subarray time delay beamforming and subarray digital beamforming the IBW is limited by the shape of the beam pattern of the subarray. As the frequency is varied from the operational center frequency, the subarray pattern squints and allows the grating lobes from the backend AF to appear in the pattern. In order to mitigate this effect, a subarray pattern that provides a windowing effect similar to a filter is required. This can be accomplished using overlapped subarrays.

Figure 3.13 shows an example of an overlapped subarray ESA architecture. Adjacent subarrays are used to form an overlapped subarray AF. This array factor provides the windowing effect to reduce grating lobes at off-tune frequencies. Ideally, a distribution would provide a window-like spatial pattern distribution as shown in Figure 3.14. The overlapped subarray AF can be expressed as

$$AF_{OSA}(\theta) = \sum_{r=1}^{R} A_{r,p} e^{j\left(\frac{2\pi}{\lambda}x_{r,p}\sin\theta - \frac{2\pi}{\lambda_o}x_{r,p}\sin\theta_o\right)} + \sum_{p=1}^{R} B_{r,p+1} \cdot e^{j\left(\frac{2\pi}{\lambda}x_{r,p+1}\sin\theta - \frac{2\pi}{\lambda_o}x_{r,p+1}\sin\theta_o\right)}$$

$$(3.13)$$

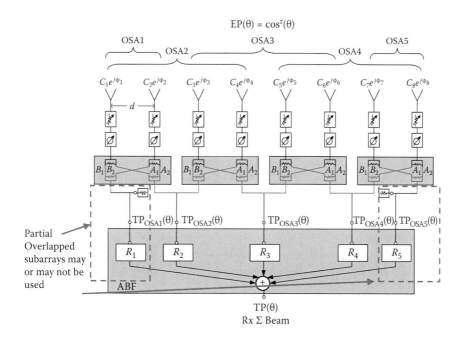

Figure 3.13 ESA architecture employing overlapped subarrays for increased IBW. (From Ulrey, J., *Sub Arrays and Beamforming*, Antenna Fundamentals Class, Northrop Grumman Electronic Systems, Baltimore, MD, 2010.)

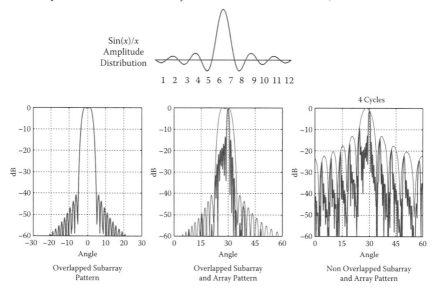

Figure 3.14 Sinx/x subarray pattern illustrates how a weighted overlapped subarray pattern distribution minimizes grating lobes for a subarray architecture.

Equation (3.13) assumes a 2:1 overlap of adjacent subarrays. The total expression for the overlapped subarray ESA can be written as

$$F(\theta) = \left(EP \cdot \sum_{r=1}^{R} A_{r,p} e^{j\left(\frac{2\pi}{\lambda}x_{r,p}\sin\theta - \frac{2\pi}{\lambda_o}x_{r,p}\sin\theta_o\right)} + \sum_{p=1}^{R} B_{r,p+1} \cdot e^{j\left(\frac{2\pi}{\lambda}x_{r,p+1}\sin\theta - \frac{2\pi}{\lambda_o}x_{r,p+1}\sin\theta_o\right)} \right)$$

$$\cdot \sum_{p=1}^{P-1} b_p \cdot e^{j\frac{2\pi}{\lambda}x_p(\sin\theta - \sin\theta_o)} \tag{3.14}$$

Equation (3.14) is similar in form to Equation (3.9). The only difference is that the overlapped subarray AF replaces that of the nonoverlapped subarray AF. The subarrays on the end of the array in Figure 3.13 have a contribution, as they form half of an overlapped subarray without a matching half. These are left out for simplicity.

Figures 3.15 and 3.16 illustrate the benefit of overlapped subarrays. The nonoverlapped subarray when tuned off of the tune frequency f_o shows grating lobes whose origin is the AF of the backend beamformer. The overlapped subarray pattern, in contrast, performs very well off of the tune frequency. As mentioned previously, the AF of the overlapped subarray spatially attenuates the AF of the backend beamformer and

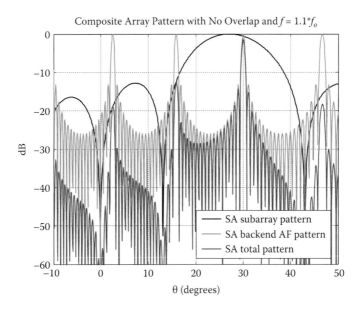

Figure 3.15 Nonoverlapped subarrays provide limited performance for off-tune operation due to backend AF grating lobes. This limits the system IBW.

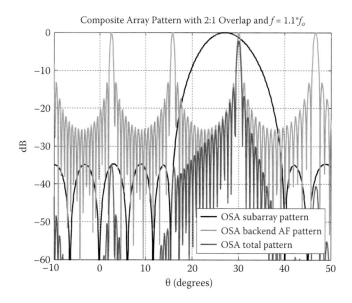

Figure 3.16 Implementation of overlapped subarrays (2:1) suppresses grating lobe for off-tune operation, thereby increasing the IBW.

provides superior IBW performance in comparison to the nonoverlapped subarray pattern.

3.5 MATLAB Program Listings

This section contains a listing of all MATLAB® programs and functions used in this chapter.

3.5.1 Compute_1D_AF.m (Function)

```
%% Function to Compute 1D AF
% Arik D. Brown
function [AF, AF_mag, AF_dB, AF_dBnorm] =...
  Compute_1D_AF(wgts,nelems,d_in,f_GHz,fo_GHz,u,uo)

lambda=11.803/f_GHz;%wavelength(in)
lambdao=11.803/fo_GHz;%wavelength at tune freq(in)

k=2*pi/lambda;%rad/in
ko=2*pi/lambdao;%rad/in

AF=zeros(1,length(u));
```

```
for ii=1:nelems
   AF = AF+wgts(ii)*exp(1j*(ii-(nelems+1)/2)*d_in*(k*u-
      ko*uo));
end
[AF_mag AF_dB AF_dBnorm] = process_vector(AF);
```

3.5.2   Compute_1D_EP.m (Function)

```
%% Function to Compute 1D EP
% Arik D. Brown

function [EP, EP_mag, EP_dB, EP_dBnorm] =...
      Compute_1D_EP(theta_deg,EF)

EP=zeros(size(theta_deg));

EP=(cosd(theta_deg).^(EF/2));%Volts
[EP_mag, EP_dB, EP_dBnorm] = process_vector(EP);
```

3.5.3   Compute_1D_PAT (Function)

```
%% Function to Compute 1D PAT
% Arik D. Brown

function [PAT, PAT_mag, PAT_dB, PAT_dBnorm] =...
      Compute_1D_PAT(EP,AF)

PAT=zeros(size(AF));

PAT=EP.*AF;
[PAT_mag PAT_dB PAT_dBnorm] =...
  process_vector(PAT);
```

3.5.4   process_vector.m (Function)

```
function[vectormag,vectordB,vectordBnorm] = process_
vector(vector)
vectormag=abs(vector);
vectordB=20*log10(vectormag+eps);
vectordBnorm=20*log10((vectormag+eps)/max(vectormag));
```

3.5.5   Taylor.m (Function)

```
%% Code to Generate Taylor Weights
% Arik D. Brown
% Original Code Author: F. W. Hopwood

Function [wgt] = Taylor(points,sll,nbar)
```

```
r = 10^(abs(sll)/20);
a = log(r+(r*r-1)^0.5) / pi;
sigma2 = nbar^2/(a*a+(nbar-0.5)^2);

%--Compute Fm, the Fourier coefficients of the weight set
for m=1:(nbar-1)
 for n=1:(nbar-1)
  f(n,1)=1-m*m/sigma2/(a*a+(n-0.5)*(n-0.5));
  if n ~= m
    f(n,2)=1/(1-m*m/n/n);
  end
  if n==m
    f(n,2)=1;
  end
 end
 g(1,1)=f(1,1);
 g(1,2)=f(1,2);

 for n=2:(nbar-1)
  g(n,1)=g(n-1,1)*f(n,1);
  g(n,2)=g(n-1,2)*f(n,2);
 end
 F(m)=((-1)^(m+1))/2*g(nbar-1,1)*g(nbar-1,2);
end

jj = [1:points]';
xx = (jj-1+0.5)/points - 1/2; %-- column vector
W = ones(size(jj)); %-- column vector
mm = [1:nbar-1]; %-- row vector
W = W + 2*cos(2*pi*xx*mm)*F';

WPK = 1 + 2*sum(F);
    wgt = W / WPK;
```

3.5.6   Subarray1D.m

```
%% Code to compute subarrayed architecture pattern
% Arik D. Brown

%% Input Parameters
IBWratio=1.1;%IBWratio -> f=fo
fo=4;%GHz Tune Frequency
f=IBWratio*fo;%GHz Operating Frequency

lambda=11.803/f;%inches
lambdao=11.803/fo;%inches

d=lambdao/2;%inches

theta=linspace(-90,90,721);%deg
thetao=30;%deg
```

```
u=sind(theta);
uo=sind(thetao);

SA.nelems=4;%Number of elements in Subarray
AF.nelems=4;%Number of elements in Backend AF

EF=1.5;

SA.wgts=ones(1,SA.nelems);
AF.wgts=ones(1,AF.nelems);

plotfigs.flags.EP=1;
plotfigs.flags.SA=1;
plotfigs.flags.AF=1;
plotfigs.flags.PAT=1;
plotfigs.flags.ALL=1;

plotfigs.axis.xlims=[-90 90];
plotfigs.axis.ylims=[-40 0];

%% Compute Pattern

% Element Pattern
[EP, EP_mag, EP_dB, EP_dBnorm] = Compute_1D_EP(theta,EF);

% Subarray AF
%!!! To simulate an array of elements without phase shifters
input 0 for
%the last input to Compute_1D_AF
[SA.AF, SA.AF_mag, SA.AF_dB, SA.AF_dBnorm] =...
  Compute_1D_AF(SA.wgts,SA.nelems,d,f,fo,u,uo);

%Backend AF
[AF.AF, AF.AF_mag, AF.AF_dB, AF.AF_dBnorm] =...
  Compute_1D_AF(AF.wgts,AF.nelems,SA.nelems*d,f,f,u,uo);

%Pattern = Element Pattern x Subarray AF Pattern x AF
           Pattern
PAT=EP.*SA.AF.*AF.AF;
[PAT_mag,PAT_dB,PAT_dBnorm] = process_vector(PAT);

SA.scanvalue.afsa=SA.AF_dBnorm(u==uo);
EPscanvalue=EP_dBnorm(u==uo);

patnorm.sa=SA.scanvalue.afsa+EPscanvalue;
%% Plot Patterns

if plotfigs.flags.EP == 1
   %Plot Pattern in dB, Unnormalized
   figure,clf
```

```
    set(gcf,'DefaultLineLineWidth',1.5)
    set(gcf,'DefaultTextFontSize',12,'DefaultTextFontWeight',
    'bold')
    plot(theta,EP_dBnorm),hold
    grid
    axis([plotfigs.axis.xlims plotfigs.axis.ylims])
    title(['Element Pattern'],'FontSize',14,'FontWeight',
    'bold')
    xlabel('\theta (degrees)','FontSize',12,'FontWeight',
    'bold')
    ylabel('dB','FontSize',12,'FontWeight','bold')
    set(gca,'FontSize',12,'FontWeight','bold')
    set(gcf,'color','white')
end

if plotfigs.flags.SA==1
    %Plot Pattern in dB, Unnormalized
    figure,clf
    set(gcf,'DefaultLineLineWidth',1.5)
    set(gcf,'DefaultTextFontSize',12,'DefaultTextFontWeight',
    'bold')
    plot(theta,SA.AF_dBnorm),hold
    grid
    axis([plotfigs.axis.xlims plotfigs.axis.ylims])
    title(['Subarray Pattern'],'FontSize',14,'FontWeight',
    'bold')
    xlabel('\theta (degrees)','FontSize',12,'FontWeight',
    'bold')
    ylabel('dB','FontSize',12,'FontWeight','bold')
    set(gca,'FontSize',12,'FontWeight','bold')
    set(gcf,'color','white')
end

if plotfigs.flags.AF==1
    %Plot Pattern in dB, Unnormalized
    figure,clf
    set(gcf,'DefaultLineLineWidth',1.5)
    set(gcf,'DefaultTextFontSize',12,'DefaultTextFontWeight',
    'bold')
    plot(theta,AF.AF_dBnorm),hold
    grid
    axis([plotfigs.axis.xlims plotfigs.axis.ylims])
    title(['Array Pattern'],'FontSize',14,'FontWeight','bold')
    xlabel('\theta (degrees)','FontSize',12,'FontWeight',
    'bold')
    ylabel('dB','FontSize',12,'FontWeight','bold')
    set(gca,'FontSize',12,'FontWeight','bold')
    set(gcf,'color','white')
end
```

```matlab
if plotfigs.flags.PAT==1
    %Plot Pattern in dB, Unnormalized
    figure 4),clf
    set(gcf,'DefaultLineLineWidth',1.5)
    set(gcf,'DefaultTextFontSize',12,'DefaultTextFontWeight',
    'bold')
    plot(theta,PAT_dBnorm+patnorm.sa),hold
    grid
    axis([plotfigs.axis.xlims plotfigs.axis.ylims])
    title(['Array Pattern'],'FontSize',14,'FontWeight','bold')
    xlabel('\theta (degrees)','FontSize',12,'FontWeight',
    'bold')
    ylabel('dB','FontSize',12,'FontWeight','bold')
    set(gca,'FontSize',12,'FontWeight','bold')
    set(gcf,'color','white')
    set(gca,'Position',[0.13 0.11 0.775 0.815])
end

if plotfigs.flags.ALL==1
    %Plot Pattern in dB, Unnormalized
    figure 5),clf
    set(gcf,'DefaultLineLineWidth',1.5)
    set(gcf,'DefaultTextFontSize',12,'DefaultTextFontWeight',
    'bold')
    plot(theta,EP_dBnorm,'color',[0 0 0]),hold
    plot(theta,SA.AF_dBnorm,'color',[0 .7 0])
    plot(theta,AF.AF_dBnorm,'color',[.7 0 1])
    plot(theta,PAT_dBnorm+patnorm.sa,'color',[0 0 1])
    grid
    axis([plotfigs.axis.xlims plotfigs.axis.ylims])
    title(['Composite Array Pattern, f =
    ',num2str(IBWratio),'*f_{o}'],...
        'FontSize',14,'FontWeight','bold')
    xlabel('\theta (degrees)','FontSize',12,'FontWeight',
    'bold')
    ylabel('dB','FontSize',12,'FontWeight','bold')
    set(gca,'FontSize',12,'FontWeight','bold')
    set(gcf,'color','white')
    legend('EP','SA.AF','AF','Pattern=EP*SA.AF*AF')
end
```

3.5.7　*Subarray1D_DBF.m*

```matlab
%% Code to compute subarrayed architecture pattern
% Arik D. Brown

%% Input Parameters

IBWratio=1.;%IBWratio -> f=fo
```

```
fo=4;%GHz Tune Frequency
f=IBWratio*fo;%GHz Operating Frequency

lambda=11.803/f;%inches
lambdao=11.803/fo;%inches

d=lambdao/2;%inches

theta=linspace(-90,90,721);%deg
thetao1=-6;%deg
thetao2=0;%deg
thetao3=6;%deg
u=sind(theta);
uo1=sind(thetao1);
uo2=sind(thetao2);
uo3=sind(thetao3);

SA.nelems=4;%Number of elements in Subarray
AF.nelems=4;%Number of elements in Backend AF

EF=1.5;

SA.wgts=ones(1,SA.nelems);
AF.wgts=ones(1,AF.nelems);

plotfigs.flags.EP=1;
plotfigs.flags.SA=1;
plotfigs.flags.AF=1;
plotfigs.flags.PATs=1;

plotfigs.axis.xlims=[-90 90];
plotfigs.axis.ylims=[-40 0];

%% Compute Pattern

% Element Pattern
[EP, EP_mag, EP_dB, EP_dBnorm] = Compute_1D_EP(theta,EF);

% Subarray AF
%!!! To simulate an array of elements without phase shifters
input 0 for
%the last input to Compute_1D_AF
[SA.AF, SA.AF_mag, SA.AF_dB, SA.AF_dBnorm] =...
  Compute_1D_AF(SA.wgts,SA.nelems,d,f,fo,u,uo2);

%Backend AFs for DBF
%%Beam 1
[AF.AF1, AF.AF1_mag, AF.AF1_dB, AF.AF1_dBnorm] =...
  Compute_1D_AF(AF.wgts,AF.nelems,SA.nelems*d,f,f,u,uo1);
```

```
%%Beam 2
[AF.AF2, AF.AF2_mag, AF.AF2_dB, AF.AF2_dBnorm] =...
  Compute_1D_AF(AF.wgts,AF.nelems,SA.nelems*d,f,f,u,uo2);

%%Beam 3
[AF.AF3, AF.AF3_mag, AF.AF3_dB, AF.AF3_dBnorm] =...
  Compute_1D_AF(AF.wgts,AF.nelems,SA.nelems*d,f,f,u,uo3);

%Pattern = Element Pattern x Subarray AF Pattern x AF Pattern
PAT1=EP.*SA.AF.*AF.AF1;
[PAT1_mag,PAT1_dB,PAT1_dBnorm] = process_vector(PAT1);

PAT2=EP.*SA.AF.*AF.AF2;
[PAT2_mag,PAT2_dB,PAT2_dBnorm] = process_vector(PAT2);

PAT3=EP.*SA.AF.*AF.AF3;
[PAT3_mag,PAT3_dB,PAT3_dBnorm] = process_vector(PAT3);

SA.scanvalue.afsa1=SA.AF_dBnorm(u==uo1);
SA.scanvalue.afsa2=SA.AF_dBnorm(u==uo2);
SA.scanvalue.afsa3=SA.AF_dBnorm(u==uo3);
EPscanvalue=EP_dBnorm(u==uo);

patnorm.sa1=SA.scanvalue.afsa1+EPscanvalue;
patnorm.sa2=SA.scanvalue.afsa2+EPscanvalue;
patnorm.sa3=SA.scanvalue.afsa3+EPscanvalue;

%% Plot Patterns
if plotfigs.flags.EP == 1
   %Plot Pattern in dB, Unnormalized
   figure,clf
   set(gcf,'DefaultLineLineWidth',1.5)
   set(gcf,'DefaultTextFontSize',12,'DefaultTextFontWeight',
   'bold')
   plot(theta,EP_dBnorm),hold
   grid
   axis([plotfigs.axis.xlims plotfigs.axis.ylims])
   title(['Element Pattern'],'FontSize',14,'FontWeight',
   'bold')
   xlabel('\theta (degrees)','FontSize',12,'FontWeight',
   'bold')
   ylabel('dB','FontSize',12,'FontWeight','bold')
   set(gca,'FontSize',12,'FontWeight','bold')
   set(gcf,'color','white')
end

if plotfigs.flags.SA==1
   %Plot Pattern in dB, Unnormalized
```

```
      figure,clf
      set(gcf,'DefaultLineLineWidth',1.5)
      set(gcf,'DefaultTextFontSize',12,'DefaultTextFontWeight',
      'bold')
      plot(theta,SA.AF_dBnorm),hold
      grid
      axis([plotfigs.axis.xlims plotfigs.axis.ylims])
      title(['Subarray Pattern'],'FontSize',14,'FontWeight',
      'bold')
      xlabel('\theta (degrees)','FontSize',12,'FontWeight',
      'bold')
      ylabel('dB','FontSize',12,'FontWeight','bold')
      set(gca,'FontSize',12,'FontWeight','bold')
      set(gcf,'color','white')
end

if plotfigs.flags.AF==1
      %Plot Pattern in dB, Unnormalized
      figure,clf
      set(gcf,'DefaultLineLineWidth',1.5)
      set(gcf,'DefaultTextFontSize',12,'DefaultTextFontWeight',
      'bold')
      plot(theta,AF.AF_dBnorm),hold
      grid
      axis([plotfigs.axis.xlims plotfigs.axis.ylims])
      title(['Array Pattern'],'FontSize',14,'FontWeight', 'bold')
      xlabel('\theta (degrees)','FontSize',12,'FontWeight',
      'bold')
      ylabel('dB','FontSize',12,'FontWeight','bold')
      set(gca,'FontSize',12,'FontWeight','bold')
      set(gcf,'color','white')
end

if plotfigs.flags.PATs==1
      %Plot Pattern in dB, Unnormalized
      figure 5),clf
      set(gcf,'DefaultLineLineWidth',1.5)
      set(gcf,'DefaultTextFontSize',12,'DefaultTextFontWeight',
      'bold')
      plot(theta,SA.AF_dBnorm,'color',[0 0 0]),hold
      plot(theta,PAT1_dBnorm+patnorm.sa1,'--','color',[.7 0 1])
      plot(theta,PAT2_dBnorm+patnorm.sa2,'-','color',[0 0 1])
      plot(theta,PAT3_dBnorm+patnorm.sa3,'--','color',[0 .7 0])
      grid
      axis([plotfigs.axis.xlims plotfigs.axis.ylims])
      title(['Composite Array Pattern'],'FontSize',14,
      'FontWeight', 'bold')
      xlabel('\theta (degrees)','FontSize',12,'FontWeight',
      'bold')
```

```
    ylabel('dB','FontSize',12,'FontWeight','bold')
    set(gca,'FontSize',12,'FontWeight','bold')
    set(gcf,'color','white')
    legend('SA.AF','Beam1','Beam2','Beam3')
end
```

3.5.8   Subarray1D_Overlapped.m

```
%% Code to compute subarrayed architecture pattern
% Arik D. Brown

%% Input Parameters
IBWratio=1.1;%IBWratio -> f=fo
fo=4;%GHz Tune Frequency
f=IBWratio*fo;%GHz Operating Frequency

lambda=11.803/f;%inches
lambdao=11.803/fo;%inches

d=lambdao/2;%inches

theta=linspace(-10,50,721);%deg
thetao=30;%deg
u=sind(theta);
uo=sind(thetao);

OSA.ratio=2;

SA.nelems=8;%Number of elements in Subarray
SA.nsas=20;%Number of Subarrays combined in Backend AF

OSA.nelems=OSA.ratio*SA.nelems;%Number of elements in
 Overlapped Subarray(OSA)
OSA.nsas=SA.nsas-(OSA.ratio-1);%Number of OSAs combined in
 Backend OSA AF
EF=1*1.5;

SA.wgts.elems=ones(1,SA.nelems);
% SA.wgts.elems=Taylor(SA.nelems,35,6);
SA.wgts.sas=ones(1,SA.nsas);

% OSA.wgts.elems=ones(1,OSA.nelems);
OSA.wgts.elems=Taylor(OSA.nelems,35,6);
OSA.wgts.sas=ones(1,OSA.nsas);

plotfigs.flags.EP=1;
plotfigs.flags.SA=1;
plotfigs.flags.AF=1;
```

```
plotfigs.flags.PATs=1;
plotfigs.flags.ALL=1;

plotfigs.axis.xlims=[min(theta) max(theta)];
plotfigs.axis.ylims=[-60 0];

%% Compute Pattern

% Element Pattern
[EP, EP_mag, EP_dB, EP_dBnorm] = Compute_1D_EP(theta,EF);

% Subarray and Overlapped Subarrays
% Subarray AF
%!!! To simulate an array of elements without phase shifters
input 0 for
%the last input to Compute_1D_AF
[SA.AFsa, SA.AFsa_mag, SA.AFsa_dB, SA.AFsa_dBnorm] =...
  Compute_1D_AF(SA.wgts.elems,SA.nelems,d,f,fo,u,uo);
%OSA AF
[OSA.AFsa, OSA.AFsa_mag, OSA.AFsa_dB, OSA.AFsa_dBnorm] =...
  Compute_1D_AF(OSA.wgts.elems,OSA.nelems,d,f,fo,u,uo);

%Backend AFs for Subarray and OSA
%Subarray Beamforming
[SA.AF, SA.AF_mag, SA.AF_dB, SA.AF_dBnorm] =...
  Compute_1D_AF(SA.wgts.sas,SA.nsas,SA.nelems*d,f,f,u,uo);
%Overlapped Subarray Beamforming
[OSA.AF, OSA.AF_mag, OSA.AF_dB, OSA.AF_dBnorm] =...
  Compute_1D_AF(OSA.wgts.sas,OSA.nsas,SA.nelems*d,f,f,u,uo);

%Pattern = Element Pattern x Subarray AF Pattern x AF
Pattern
SA.PAT=EP.*SA.AFsa.*SA.AF;
[SA.PAT_mag,SA.PAT_dB,SA.PAT_dBnorm] = process_vector(SA.PAT);
OSA.PAT=EP.*OSA.AFsa.*OSA.AF;
[OSA.PAT_mag,OSA.PAT_dB,OSA.PAT_dBnorm] = process_
 vector(OSA.PAT);

SA.scanvalue.afsa=SA.AFsa_dBnorm(u==uo);
OSA.scanvalue.afsa=OSA.AFsa_dBnorm(u==uo);
EPscanvalue=EP_dBnorm(u==uo);

patnorm.sa=SA.scanvalue.afsa+EPscanvalue;
patnorm.osa=OSA.scanvalue.afsa+EPscanvalue;

%% Plot Patterns

if plotfigs.flags.EP == 1
    %Plot Pattern in dB, Normalized
    figure,clf
    set(gcf,'DefaultLineLineWidth',1.5)
```

```
    set(gcf,'DefaultTextFontSize',12,'DefaultTextFontWeight',
    'bold')
    plot(theta,EP_dBnorm),hold
    grid
    axis([plotfigs.axis.xlims plotfigs.axis.ylims])
    title(['Element Pattern'],'FontSize',14,'FontWeight',
    'bold')
    xlabel('\theta (degrees)','FontSize',12,'FontWeight',
    'bold')
    ylabel('dB','FontSize',12,'FontWeight','bold')
    set(gca,'FontSize',12,'FontWeight','bold')
    set(gcf,'color','white')
end
if plotfigs.flags.SA==1
    %Plot SA Pattern in dB, Normalized
    figure,clf
    set(gcf,'DefaultLineLineWidth',1.5)
    set(gcf,'DefaultTextFontSize',12,'DefaultTextFontWeight',
    'bold')
    plot(theta,SA.AFsa_dBnorm),hold
    plot(theta,OSA.AFsa_dBnorm,'--')
    grid
    axis([plotfigs.axis.xlims plotfigs.axis.ylims])
    title(['Subarray and OSA AF Patterns'],'FontSize',14,
    'FontWeight','bold')
    xlabel('\theta (degrees)','FontSize',12,'FontWeight',
    'bold')
    ylabel('dB','FontSize',12,'FontWeight','bold')
    set(gca,'FontSize',12,'FontWeight','bold')
    set(gcf,'color','white')
end
if plotfigs.flags.AF==1
    %Plot AF in dB, Normalized
    figure,clf
    set(gcf,'DefaultLineLineWidth',1.5)
    set(gcf,'DefaultTextFontSize',12,'DefaultTextFontWeight',
    'bold')
    plot(theta,SA.AF_dBnorm),hold
    plot(theta,OSA.AF_dBnorm,'--')
    grid
    axis([plotfigs.axis.xlims plotfigs.axis.ylims])
    title(['AF Pattern'],'FontSize',14,'FontWeight','bold')
    xlabel('\theta (degrees)','FontSize',12,'FontWeight','
    bold')
    ylabel('dB','FontSize',12,'FontWeight','bold')
    set(gca,'FontSize',12,'FontWeight','bold')
    set(gcf,'color','white')
end
```

```
if plotfigs.flags.PATs==1
   %Plot Pattern in dB, Normalized
   figure,clf
   set(gcf,'DefaultLineLineWidth',1.5)
   set(gcf,'DefaultTextFontSize',12,'DefaultTextFontWeight',
   'bold')
   plot(theta,SA.PAT_dBnorm,'-','color',[.7 0 1]),hold
   plot(theta,OSA.PAT_dBnorm,'--','color',[0 0 1])
   grid
   axis([plotfigs.axis.xlims plotfigs.axis.ylims])
   title(['Composite Array Pattern with ',num2str(OSA.
   ratio),...
      ': 1 Overlap and f = ',num2str(IBWratio),'*f_{o}'],...
      'FontSize',14,'FontWeight','bold')
   xlabel('\theta (degrees)','FontSize',12,'FontWeight',
   'bold')
   ylabel('dB','FontSize',12,'FontWeight','bold')
   set(gca,'FontSize',12,'FontWeight','bold')
   set(gcf,'color','white')
   legend('No Overlap Pattern',[num2str(OSA.ratio),': 1 OSA
   Pattern'])
end

if plotfigs.flags.ALL==1
   %Plot Pattern in dB, Normalized
   figure,clf
   set(gcf,'DefaultLineLineWidth',1.5)
   set(gcf,'DefaultTextFontSize',12,'DefaultTextFontWeight',
   'bold')
   plot(theta,OSA.AFsa_dBnorm,'k-'),hold
   plot(theta,OSA.AF_dBnorm,'-','color',[0 0.7 0])
   plot(theta,OSA.PAT_dBnorm+patnorm.osa,'-','color',[0 0 1])
   grid
   axis([plotfigs.axis.xlims plotfigs.axis.ylims])
   title(['Composite Array Pattern with ',num2str(OSA.
   ratio),...
      ': 1 Overlap and f = ',num2str(IBWratio),'*f_{o}'],...
      'FontSize',14,'FontWeight','bold')
   xlabel('\theta (degrees)','FontSize',12,'FontWeight',
   'bold')
   ylabel('dB','FontSize',12,'FontWeight','bold')
   set(gca,'FontSize',12,'FontWeight','bold')
   set(gcf,'color','white')
   legend('OSA Subarray Pattern','OSA Backend AF
   Pattern','OSA Total Pattern')

   %Plot Pattern in dB, Normalized
   figure,clf
   set(gcf,'DefaultLineLineWidth',1.5)
```

```
    set(gcf,'DefaultTextFontSize',12,'DefaultTextFontWeight',
    'bold')
    plot(theta,SA.AFsa_dBnorm,'k-'),hold
    plot(theta,SA.AF_dBnorm,'-','color',[0 0.7 0])
    plot(theta,SA.PAT_dBnorm+patnorm.sa,'-','color',[0 0 1])
    grid
    axis([plotfigs.axis.xlims plotfigs.axis.ylims])
    title(['Composite Array Pattern with No Overlap and f =
    ',num2str(IBWratio),'*f_{o}'],...
       'FontSize',14,'FontWeight','bold')
    xlabel('\theta (degrees)','FontSize',12,'FontWeight',
    'bold')
    ylabel('dB','FontSize',12,'FontWeight','bold')
    set(gca,'FontSize',12,'FontWeight','bold')
    set(gcf,'color','white')
    legend('SA Subarray Pattern','SA Backend AF Pattern','SA
    Total Pattern')
end
```

References

Skolnik, Merrill. *Radar Handbook*, 2nd ed. McGraw-Hill, New York, NY, 1990.

Ulrey, Jonathan. *Sub Arrays and Beamforming*. Antenna Fundamentals Class. Baltimore, MD: Northrop Grumman Electronic Systems, 2010.

Wojtowicz, John. *Instantaneous Bandwidth Analysis and Subarrayed Antennas*. Antenna Systems Class. Baltimore, MD: Northrop Grumman Electronic Systems, April 2000.

chapter four

Pattern Optimization

Daniel Boeringer
Northrop Grumman Electronic Systems

Contents

4.1   Introduction

One of the virtues of phased arrays is the ability to create almost arbitrary far field pattern characteristics, by tailoring the amplitude and phase characteristics of the aperture, as enabled by the phase shifters and attenuators at each element. However, it is not always obvious how to achieve a desired far field pattern, or even to know if a desired far field pattern is achievable with a given aperture. When available, analytical techniques often provide solutions with demonstrable optimality, but can often require significant mathematical derivation, such as computation of derivatives. This can be impractical when the desired characteristics are especially nonlinear.

On the other hand, stochastic methods, although not guaranteed to converge to the best possible solution, are easy to apply and can give a rapid "engineering" answer as to the achievability of a desired far field pattern characteristic. This chapter will provide code and detailed examples of optimization using two variants of common stochastic optimizers, a particle swarm optimizer and a genetic algorithm. More important than the particular optimizers chosen here is understanding their application to phased array pattern synthesis, which can facilitate the application of this or other optimizers to problems of interest to the reader.

4.2   Stochastic Pattern Optimization Overview

As shown in Figure 4.1, stochastic optimizers interact with a cost function to optimize the ultimate solution. For a given trial solution, a cost function returns a numerical value that characterizes how well the trial solution performs. By repeatedly evaluating the cost function with different trial solutions, the optimizer adjusts the trial solutions and attempts to converge to the best solution, defined by the solution returning the lowest cost. The designer then reviews the outcome, and may then adjust the cost function or run the optimizer for more iterations to further refine the solution.

4.2.1   Particle Swarm Optimization Overview

Particle swarm optimization is very easy to understand and implement, with little computational bookkeeping. In the code provided, the code devoted to optimization only comprises about 30 lines. For the simultaneous optimization of N variables, a collection or swarm of

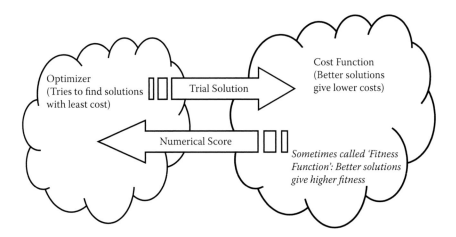

Figure 4.1 Optimizer repeatedly interacts with cost function to optimize solution.

particles is defined, where each particle is assigned a random position in the N-dimensional problem space so that each particle's position corresponds to a candidate solution to the optimization problem. Each of these particle positions is scored to obtain a scalar cost based on how well it solves the problem. These particles are then flown through the N-dimensional problem space subject to both deterministic and stochastic update rules to new positions, which are subsequently scored. As the particles traverse the problem hyperspace, each particle remembers it is the personal best position that it has ever found, called its local best. Each particle also knows the best position found by any particle in the swarm, called the global best. On successive iterations, particles are "tugged" toward these prior best solutions with linear spring-like attraction forces. Overshoot and undershoot combined with stochastic adjustment explore regions throughout the problem hyperspace, eventually settling down near a good solution. A detailed discussion of the particle swarm optimizer used in this chapter is provided in Boeringer and Werner (2004).

4.2.2   Genetic Algorithm Overview

Inspired by Darwin's theories of evolution and the concept of survival of the fittest, genetic algorithms use processes analogous to genetic recombination and mutation to promote the evolution of a population that best satisfies a predefined goal. In the selective breeding or crossover process, fit individuals are chosen to produce more offspring than less fit individuals, which tends to homogenize the population and improve the average result as the algorithm progresses. Subsequent mutations of the offspring add diversity to the population and explore new areas of the parameter search space. The algorithm starts by generating an initial population of random candidate solutions. Each individual in the population is then awarded a score based on its performance. The individuals with the best scores are chosen to be parents, which are cut and spliced together to make children. To add some diversity to the population, some random mutations are applied. These children are scored, with the best performers likely to be parents in the next generation. At some point the process is terminated and the best scoring individual in the population is taken as the final answer. A detailed discussion of the genetic algorithm used in this chapter is provided in Boeringer and Werner (2004).

4.3   Pattern Optimization Implementation

Figure 4.2 illustrates how a numerical vector generated by either the particle swarm optimizer or the genetic algorithm is converted to aperture weights and scored against the desired far field pattern. Both optimizers

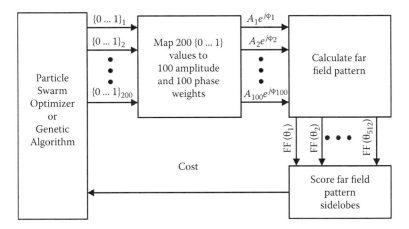

Figure 4.2 Numerical vector generated by optimizer is converted to aperture weights and scored against the desired far field pattern.

generate trial solutions as a set of numbers between 0 and 1, which are then linearly mapped to the amplitude and phase at each element. The MATLAB® codes in this chapter implement this block diagram and are explained further below.

4.3.1   Cost Function Construction

A typical cost function goal for the optimization of array weights for a linear array is described in Figure 4.3, where cost is defined as the excess sidelobe power outside specified upper and lower goals. The lower the cost, the better the solution. For a given far field pattern, each pattern point that lies outside the specified limits contributes a value to the cost function equal to the power difference between the goal and the far field pattern. The cost function given in simpleCostFunction.m implements this cost function. The inputs to this subroutine include a trial set of array weights, the far field amplitude from individual array elements as a function of far field angle, the desired sidelobe limits, and some flags that govern the cost function evaluation. One flag forces the aperture weights to be symmetric across the aperture, if set. The FFT function is used to evaluate the far field pattern, which implicitly assumes that the phased array elements are spaced at half wavelength, and the pattern power outside the sidelobe limits is tallied. If the flag to exempt the main beam from the sidelobe constraint is selected, then a simple numerical differentiation is used to identify the main beam and exclude those points from the tally of pattern power outside the sidelobe limits. This is convenient because it does not require the user to know the exact main beam beamwidth before the

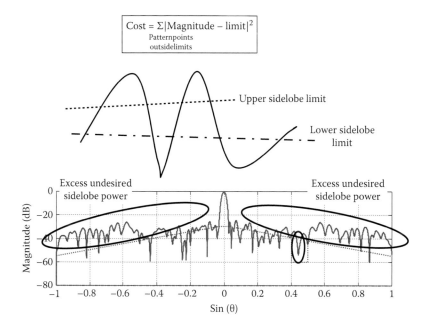

Figure 4.3 Cost is the excess sidelobe power outside specified upper and lower goals.

optimization and adjust the sidelobe targets accordingly. The output is a single number, the logarithm of the total pattern power outside the side-lobe limits. The logarithm allows the user to see finer improvements as the optimizer converges. This single number is the cost function's score of the trial solution.

4.3.2   Particle Swarm Optimization Examples

Three examples of particle swarm optimization are given, to illustrate how different types of problems can be handled with substantially the same code. The code simplePS_AmpTaperEx.m produced the amplitude-only synthesis of −30 dB sidelobes with a −60 dB notch using particle swarm optimization, shown in Figure 4.4. The parameters at the top of the code specify a 100-element phased array, with an element factor for each element that varies as $\cos(q)^{1.2}$. A range (minimum and maximum) of permissible values for each amplitude and phase value is specified; for amplitude-only optimization, the range of amplitude values is 0 to 1, while the range for phase is 0 to 0 to enforce constant phase. For this amplitude-only synthesis problem, a flag is set constraining the output distribution to be symmetrical. The sidelobe upper and lower goals, as shown in Figure 4.4, are

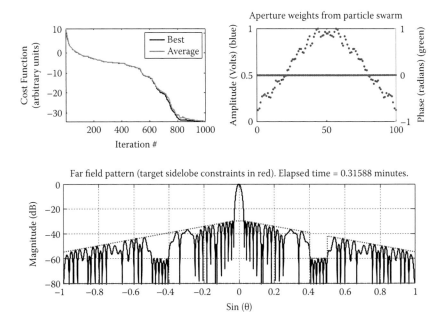

Figure 4.4 Amplitude-only synthesis of −30 dB sidelobes with a −60 dB notch using particle swarm optimization.

specified as lists of points that are interpolated in the cost function evaluation. The rest of the code performs the optimization as described in Section 4.2.1 and Figure 4.2, and displays the results. As shown in the bottom of Figure 4.4, the specifications are well met, including the deep notch. The convergence of the optimizer as a function of iteration is shown in the top left of the figure, while the optimized amplitude and phase weights are shown in the top right.

A simple modification to the inputs provides phase-only optimization. The code simplePS_PhaseOnlyEx.m produced the phase-only synthesis of −30 dB sidelobes with a −60 dB notch goal using particle swarm optimization shown in Figure 4.5. The modified specifications include constraining the amplitude weights to Taylor weights and allowing the phase weights to vary between 0 and p/2. The symmetry constraint is removed for this case. With these constraints, the results are not quite as good as the amplitude-only case, but at −50 dB the notch is still deep.

A further modification of the basic code demonstrates complex (amplitude and phase) synthesis for a symmetrical flat-top beam. The code simplePS_FlattopEx produced the complex synthesis of −35 dB sidelobes with a flat-top main beam using particle swarm optimization shown in Figure 4.6. Here the amplitude is allowed to vary between 0 and 1, while the phase is

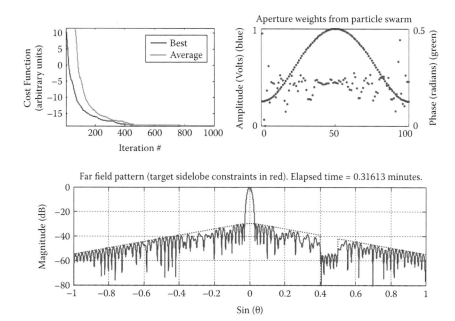

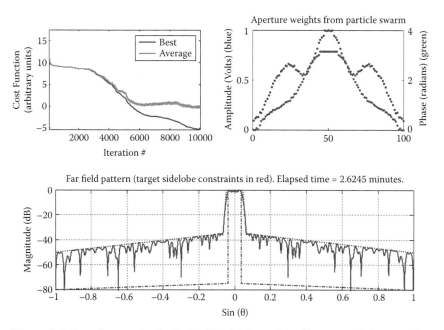

Figure 4.5 Phase-only synthesis of –30 dB sidelobes with a –60 dB notch using particle swarm optimization.

Figure 4.6 Complex synthesis of –35 dB sidelobes with a flat-top main beam using particle swarm optimization.

allowed to vary between 0 and p. The sidelobe constraints are adjusted to produce a flat-top beam, where the lower sidelobe level constraint is used to "pinch" the main beam at the flat top. For this case, the symmetry constraint is turned on, while the flag to exempt the main beam from the cost function is turned off. With the full freedom to adjust both amplitude and phase, the flat-top specification is well met; however, note that this optimization required more iterations to converge than the previous two examples.

4.3.3   Genetic Algorithm Examples

The same three examples given above are now approached with the genetic algorithm. The code simpleGA_AmpTaperEx.m has the same specifications as simplePS_AmpTaperEx.m, but implements the optimization with the genetic algorithm as described in Section 4.2.2 and Figure 4.2. The results shown in Figure 4.7 are comparable to those shown in Figure 4.4, but note that the ultimate cost function value is better for the genetic algorithm than for the particle swarm optimizer. The phase-only optimization problem is solved with the code simpleGA_PhaseOnlyEx.m, with the results shown in Figure 4.8. The complex synthesis problem is solved with the code simpleGA_FlattopEx.m, with the results shown in Figure 4.9, again

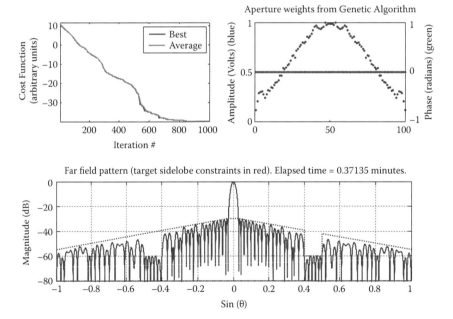

Figure 4.7 Amplitude-only synthesis of −30 dB sidelobes with a −60 dB notch using a genetic algorithm.

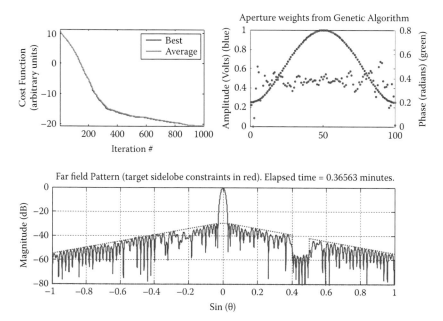

Figure 4.8 Phase-only synthesis of −30 dB sidelobes with a −60 dB notch using a genetic algorithm.

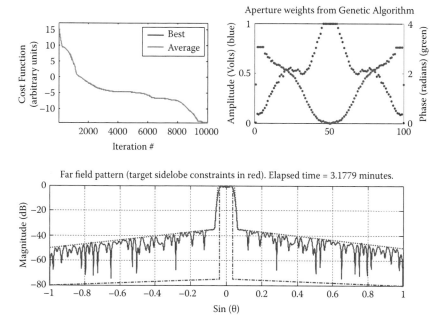

Figure 4.9 Complex synthesis of −35 dB sidelobes with a flat-top main beam using a genetic algorithm.

with a lower cost function value than for the particle swarm optimizer. It is not always the case that genetic algorithms outperform particle swarm optimizers; rather, the two algorithms take different journeys through the cost function space, and so one algorithm may outperform the other on any given problem. Having both algorithms available and interchangeable can be part of the overall design process, in addition to pattern specification and cost function tuning.

4.4 MATLAB Program and Function Listings

This section contains a listing of all MATLAB programs and functions used in this chapter.

4.4.1 simpleCostFunction.m

```
function [Score]=simplePS_CostFunction(ComplexWgts,ElementWg
tVoltage,LowerEnvelope,UpperEnvelope,AutomaticallyExemptMain
Beam,EnforceSymmetry);

Nelements=prod(size(ComplexWgts));
if EnforceSymmetry
  ComplexWgts(Nelements/2+1:Nelements)=flipud(ComplexWgts
  (1:Nelements/2));
end

NPatternPoints=prod(size(ElementWgtVoltage));
Pattern=abs(fftshift ( fft(ComplexWgts,NPatternPoints))).*El
ementWgtVoltage;
PatternNorm=Pattern/max(Pattern);
LowerExcess=max(LowerEnvelope-PatternNorm,0);
UpperExcess=max(PatternNorm-UpperEnvelope,0);

if AutomaticallyExemptMainBeam
  MaxValue=max(PatternNorm);
  IndexOfMaxValue=find(PatternNorm==MaxValue);
  Spikes=diff(sign(diff(PatternNorm)));
  NullIndices=find(Spikes==2)+1;
  PeakIndices=find(Spikes==-2)+1;
  RightNullIndices=NullIndices(find(NullIndices>IndexOfMaxVa
  lue(1)));
  LeftNullIndices=NullIndices(find(NullIndices<IndexOfMaxVa
  lue(1)));
  if isempty(RightNullIndices)
    IndexOfRightNull=NPatternPoints;
  else
    IndexOfRightNull=RightNullIndices(1);
  end
```

```
if isempty(LeftNullIndices)
  IndexOfLeftNull=1;
else
  IndexOfLeftNull=LeftNullIndices(prod(size(LeftNullIndi
  ces)));
end
LowerExcess(IndexOfLeftNull:IndexOfRightNull)=0;
UpperExcess(IndexOfLeftNull:IndexOfRightNull)=0;
end

Score=10*log10(sum(LowerExcess)+sum(UpperExcess)+eps);
```

4.4.2   *simpleGA_AmpTaperEx.m*

```
clc
close all
clear all
%%%%%%%%%%%%%%%%%%%%%%%%%%%%%%%%%%%%%%%%%
% Antenna Description
%%%%%%%%%%%%%%%%%%%%%%%%%%%%%%%%%%%%%%%%%
Nelements=100;
CosineElementFactorExponent=1.2;
%%%%%%%%%%%%%%%%%%%%%%%%%%%%%%%%%%%%%%%%%
% Excitation Constraints
%%%%%%%%%%%%%%%%%%%%%%%%%%%%%%%%%%%%%%%%%
MagMin(1:Nelements,1)=0;
MagMax(1:Nelements,1)=1;
PhsMin(1:Nelements,1)=0;
PhsMax(1:Nelements,1)=0;
EnforceSymmetry=1; % 1 for symmetric distrib, 0 for asymmetric
%%%%%%%%%%%%%%%%%%%%%%%%%%%%%%%%%%%%%%%%%
% Pattern Goals
%%%%%%%%%%%%%%%%%%%%%%%%%%%%%%%%%%%%%%%%%
UpperSidelobeGoal_u= [ -1 -.03 .03 .4 .4 .5 .5 1 ];
UpperSidelobeGoal_dB=[-55 -30 -30 -39.5 -60 -60 -42.1 -55];
LowerSidelobeGoal_u= [ -1 1];
LowerSidelobeGoal_dB=[-80 -80];
AutomaticallyExemptMainBeam=1; % 1 to exempt main beam
from goals
NPatternPoints=512;
%%%%%%%%%%%%%%%%%%%%%%%%%%%%%%%%%%%%%%%%%
% Optimization Parameters
%%%%%%%%%%%%%%%%%%%%%%%%%%%%%%%%%%%%%%%%%
Niterations=1000;
Npopulation=50;
Nmarriages=25;
TournamentEligible=10;
TournamentCompetitors=5;
```

```
Ncrossovers=2;
MutationProbability=0.04;
MutationRangeMax=0.2;
MutationRangeDecay=1.5;
Norder=2*Nelements;
Nmutations=round(MutationProbability*Norder);
%%%%%%%%%%%%%%%%%%%%%%%%%%%%%%%%%%%%%%%
% Miscellaneous Initialization
%%%%%%%%%%%%%%%%%%%%%%%%%%%%%%%%%%%%%%%
MagRange=MagMax-MagMin;
PhsRange=PhsMax-PhsMin;
SineTheta=[-1:2/NPatternPoints:1-2/NPatternPoints];
CosineTheta=sqrt(1-SineTheta.^2);
LowerEnvelope_dB=interp1(LowerSidelobeGoal_
u+[0:size(LowerSidelobeGoal_u,2)-1]*eps, LowerSidelobeGoal_
dB, SineTheta);
UpperEnvelope_dB=interp1(UpperSidelobeGoal_
u+[0:size(UpperSidelobeGoal_u,2)-1]*eps, UpperSidelobeGoal_
dB, SineTheta);
LowerEnvelope=10.^(LowerEnvelope_dB/20)';
UpperEnvelope=10.^(UpperEnvelope_dB/20)';
ElementWgtPower=CosineTheta.^CosineElementFactorExponent;
ElementWgtVoltage=sqrt(ElementWgtPower)';
tic;


%%%%%%%%%%%%%%%%%%%%%%%%%%%%%%%%%%%%%%%
% Initialize Population
%%%%%%%%%%%%%%%%%%%%%%%%%%%%%%%%%%%%%%%
CurrentGeneration=rand(Norder,Npopulation);
%%%%%%%%%%%%%%%%%%%%%%%%%%%%%%%%%%%%%%%
% Score Population
%%%%%%%%%%%%%%%%%%%%%%%%%%%%%%%%%%%%%%%
for kk=1:Npopulation
  MagWgts=CurrentGeneration(1:Nelements,kk).*MagRange+Mag
  Min;
  PhsWgts=CurrentGeneration(Nelements+1:2*Nelements,kk).*Phs
  Range+PhsMin;
  ComplexWgts=(MagWgts.*exp(i*PhsWgts));
  CurrentGenerationScores(kk)=simpleCostFunction(...
    ComplexWgts,ElementWgtVoltage,LowerEnvelope,UpperEnvel
    ope,...
    AutomaticallyExemptMainBeam,EnforceSymmetry);
end
%%%%%%%%%%%%%%%%%%%%%%%%%%%%%%%%%%%%%%%
% Get Ready to Optimize
%%%%%%%%%%%%%%%%%%%%%%%%%%%%%%%%%%%%%%%
[SortFitness,SortIndx]=sort(CurrentGenerationScores);
CurrentGeneration=CurrentGeneration(:,SortIndx);
CurrentGenerationScores=CurrentGenerationScores(SortIndx);
```

```
GlobalBestScore=CurrentGenerationScores(1);
%%%%%%%%%%%%%%%%%%%%%%%%%%%%%%%%%%%%%
% Main Optimization Loop
%%%%%%%%%%%%%%%%%%%%%%%%%%%%%%%%%%%%%
for ii=1:Niterations
  t=(ii-1)/(Niterations-1);
  MutationRange=MutationRangeMax*(exp(-
MutationRangeDecay*t)-exp(-MutationRangeDecay));
  %%%%%%%%%%%%%%%%%%%%%%%%%%%%%%%%%%%%%%
  % For Each Marriage:
  %%%%%%%%%%%%%%%%%%%%%%%%%%%%%%%%%%%%%%
  for kk=1:Nmarriages
    %%%%%%%%%%%%%%%%%%%%%%%%%%%%%%%%%%%%%%%%
    % Choose Mom
    %%%%%%%%%%%%%%%%%%%%%%%%%%%%%%%%%%%%%%%%
    PermutedIndx=randperm(TournamentEligible);
    CompetitorFitness=CurrentGeneration(PermutedIndx(1:
    TournamentCompetitors));
    WinnerIndx=find(CompetitorFitness==min(CompetitorFitn
    ess));
    Mom=CurrentGeneration(:,PermutedIndx(WinnerIndx(1)));
    %%%%%%%%%%%%%%%%%%%%%%%%%%%%%%%%%%%%%%%%
    % Choose Dad
    %%%%%%%%%%%%%%%%%%%%%%%%%%%%%%%%%%%%%%%%
    PermutedIndx=randperm(TournamentEligible);
    CompetitorFitness=CurrentGeneration(PermutedIndx(1:
    TournamentCompetitors));
    WinnerIndx=find(CompetitorFitness==min(CompetitorFitn
    ess));
    Dad=CurrentGeneration(:,PermutedIndx(WinnerIndx(1)));
    %%%%%%%%%%%%%%%%%%%%%%%%%%%%%%%%%%%%%%%%
    % Mix Mom and Dad to Make Kids
    %%%%%%%%%%%%%%%%%%%%%%%%%%%%%%%%%%%%%%%%
    Son=Dad;
    Daughter=Mom;
    PermutedIndx=randperm(Norder-1)+1;
    CrossoverPoints=sort(PermutedIndx(1:Ncrossovers));
    for CrossOverCount=1:Ncrossovers
      Transfer=Son(CrossoverPoints(CrossOverCount):Norder);

Son(CrossoverPoints(CrossOverCount):Norder)=Daughter(Crossov
erPoints(CrossOverCount):Norder);
      Daughter(CrossoverPoints(CrossOverCount):Norder)=Trans
      fer;
    end
    %%%%%%%%%%%%%%%%%%%%%%%%%%%%%%%%%%%%%%%%
    % Mutate Son
    %%%%%%%%%%%%%%%%%%%%%%%%%%%%%%%%%%%%%%%%
    PermutedIndx=randperm(Norder);
```

```
MutationDecision=zeros(Norder,1);
MutationDecision(PermutedIndx(1:Nmutations))=1;
MutatedMax=min(Son+MutationRange/2, 1);
MutatedMin=max(Son-MutationRange/2, 0);
Mutations=rand(Norder,1).*(MutatedMax-
MutatedMin)+MutatedMin;
Children(:,2*kk-1)=Son.*(1-MutationDecision)+Mutations.*
MutationDecision;
%%%%%%%%%%%%%%%%%%%%%%%%%%%%%%%%%%%%%%%
% Mutate Daughter
%%%%%%%%%%%%%%%%%%%%%%%%%%%%%%%%%%%%%%%
PermutedIndx=randperm(Norder);
MutationDecision=zeros(Norder,1);
MutationDecision(PermutedIndx(1:Nmutations))=1;
MutatedMax=min(Daughter+MutationRange/2, 1);
MutatedMin=max(Daughter-MutationRange/2, 0);
Mutations=rand(Norder,1).*(MutatedMax-
MutatedMin)+MutatedMin;
Children(:,2*kk)=Daughter.*(1-MutationDecision)+Mutation
s.*MutationDecision;
end
%%%%%%%%%%%%%%%%%%%%%%%%%%%%%%%%%%%%%%%%%
% Score Children
%%%%%%%%%%%%%%%%%%%%%%%%%%%%%%%%%%%%%%%%%
for kk=1:2*Nmarriages
  MagWgts=Children(1:Nelements,kk).*MagRange+MagMin;
  PhsWgts=Children(Nelements+1:2*Nelements,kk).*PhsRange+P
  hsMin;
  ComplexWgts=(MagWgts.*exp(i*PhsWgts));
  ChildrenScores(kk)=simpleCostFunction(...
    ComplexWgts,ElementWgtVoltage,LowerEnvelope,UpperEnvel
    ope,...
    AutomaticallyExemptMainBeam,EnforceSymmetry);
end
%%%%%%%%%%%%%%%%%%%%%%%%%%%%%%%%%%%%%%%%%%
% Survival of the Fittest
%%%%%%%%%%%%%%%%%%%%%%%%%%%%%%%%%%%%%%%%%%
CombinedGenerations=[CurrentGeneration Children];
CombinedScores=[CurrentGenerationScores ChildrenScores];
[SortFitness,SortIndx]=sort(CombinedScores);
SurvivorsIndx=SortIndx(1:Npopulation);
CurrentGeneration=CombinedGenerations(:,SurvivorsIndx);
CurrentGenerationScores=CombinedScores(SurvivorsIndx);
GlobalBestScore=CurrentGenerationScores(1);
%%%%%%%%%%%%%%%%%%%%%%%%%%%%%%%%%%%%%%%%%%
% Save Performance Results
%%%%%%%%%%%%%%%%%%%%%%%%%%%%%%%%%%%%%%%%%%
```

```
  SavePopulationAverageScore(ii)=mean(CurrentGenerationSco
  res);
  SavePopulationBestScores(ii)=GlobalBestScore;
end
GlobalBestIndividual=CurrentGeneration(:,1);

%%%%%%%%%%%%%%%%%%%%%%%%%%%%%%%%%%%%%%
% Calculate Best Pattern
%%%%%%%%%%%%%%%%%%%%%%%%%%%%%%%%%%%%%%
NPatternPointsFine=NPatternPoints*8;
SineThetaFine=[-1:2/NPatternPointsFine:1-2/
NPatternPointsFine];
CosineThetaFine=sqrt(1-SineThetaFine.^2);
ElementWgtPowerFine=CosineThetaFine.^CosineElementFactorExpo
nent;
ElementWgtVoltageFine=sqrt(ElementWgtPowerFine).';
ComplexWgts=(GlobalBestIndividual(1:Nelements).*MagRange+Mag
Min).*...
  exp(i*(GlobalBestIndividual(Nelements+1:2*Nelements).*PhsR
  ange+PhsMin));
if EnforceSymmetry
  ComplexWgts(Nelements/2+1:Nelements)=flipud(ComplexWgts(1:
  Nelements/2));
end
ComplexWgts=ComplexWgts/max(abs(ComplexWgts));
BestPattern=fftshift (fft( ComplexWgts,NPatternPointsFine)).
*ElementWgtVoltageFine;
BestPattern_dB=20*log10(abs(BestPattern)+eps);
BestPatternNorm_dB=BestPattern_dB-max(BestPattern_dB);
ElapsedTimeMinutes=toc/60;

%%%%%%%%%%%%%%%%%%%%%%%%%%%%%%%%%%%%%%%%
% Plot Results
%%%%%%%%%%%%%%%%%%%%%%%%%%%%%%%%%%%%%%%%
figure
figscale=70; figoffsetx=20; figoffsety=20;
set(gcf,'Position',[figoffsetx figoffsety round(11.25*figsca
le+figoffsetx) round(6.75*figscale+figoffsety)])
fontsize=12;
subplot(2,2,1)
set(gca,'FontSize',fontsize)
plot([1:Niterations],SavePopulationBestScores,[1:Niterations]
,SavePopulationAverageScore,'LineWidth',xlabel('Iteration #')
ylabel('Cost Function (arbitrary units)')
legend('Best','Average')
axis tight
subplot(2,2,2)
```

```
[ax,h1,h2]=plotyy([1:Nelements],abs(ComplexWgts),[1:Nelement
s],angle(ComplexWgts)-min(angle(ComplexWgts)));
set(h1,'Marker','.','LineStyle','none');
set(h2,'Marker','.','LineStyle','none');
axes(ax(1));
set(gca,'FontSize',fontsize)
ylabel('Amplitude (Volts) (blue)')
ylim([0 1])
axes(ax(2));
set(gca,'FontSize',fontsize)
ylabel('Phase (radians) (green)')
title('Aperture weights from Genetic Algorithm')
subplot(2,1,2)
set(gca,'FontSize',fontsize)
plot(SineThetaFine,BestPatternNorm_dB,'LineWidth',2);
hold on
plot(LowerSidelobeGoal_u, LowerSidelobeGoal_dB,
'r-.','LineWidth',2)
plot(UpperSidelobeGoal_u, UpperSidelobeGoal_dB,
'r:','LineWidth',2)
axis tight
ylim([-80 0])
grid on
title(['Far field Pattern (Target sidelobe constraints in
red). Elapsed time = ' num2str(ElapsedTimeMinutes) '
minutes.'])
ylabel('Magnitude (dB)')
xlabel('Sin(\theta)')
```

4.4.3   *simpleGA_FlattopEx.m*

```
clc
close all
clear all
%%%%%%%%%%%%%%%%%%%%%%%%%%%%%%%%%%%%%%%%%
% Antenna Description
%%%%%%%%%%%%%%%%%%%%%%%%%%%%%%%%%%%%%%%%%
Nelements=100;
CosineElementFactorExponent=1.2;
%%%%%%%%%%%%%%%%%%%%%%%%%%%%%%%%%%%%%%%%%
% Excitation Constraints
%%%%%%%%%%%%%%%%%%%%%%%%%%%%%%%%%%%%%%%%%
MagMin(1:Nelements,1)=0;
MagMax(1:Nelements,1)=1;
PhsMin(1:Nelements,1)=0;
PhsMax(1:Nelements,1)=pi;
EnforceSymmetry=1; % 1 for symmetric distrib, 0 for asymmetric
%%%%%%%%%%%%%%%%%%%%%%%%%%%%%%%%%%%%%%%%%
```

```
% Pattern Goals
%%%%%%%%%%%%%%%%%%%%%%%%%%%%%%%%%%%%%%%%
UpperSidelobeGoal_u= [-1 -.065 -.045 .045 .065 1 ];
UpperSidelobeGoal_dB=[-50 -35 0 0 -35 -50];
LowerSidelobeGoal_u= [ -1 -.036 -.036 .036 .036 1];
LowerSidelobeGoal_dB=[-80 -75 -1 -1 -75 -80];
AutomaticallyExemptMainBeam=0; % 1 to exempt main beam
from goals
NPatternPoints=512;
%%%%%%%%%%%%%%%%%%%%%%%%%%%%%%%%%%%%%%%%
% Optimization Parameters
%%%%%%%%%%%%%%%%%%%%%%%%%%%%%%%%%%%%%%%%
Niterations=1000*10;
Npopulation=50;
Nmarriages=25;
TournamentEligible=10;
TournamentCompetitors=5;
Ncrossovers=2;
MutationProbability=0.04;
MutationRangeMax=0.2;
MutationRangeDecay=1.5;
Norder=2*Nelements;
Nmutations=round(MutationProbability*Norder);
%%%%%%%%%%%%%%%%%%%%%%%%%%%%%%%%%%%%%%%%
% Miscellaneous Initialization
%%%%%%%%%%%%%%%%%%%%%%%%%%%%%%%%%%%%%%%%
MagRange=MagMax-MagMin;
PhsRange=PhsMax-PhsMin;
SineTheta=[-1:2/NPatternPoints:1-2/NPatternPoints];
CosineTheta=sqrt(1-SineTheta.^2);
LowerEnvelope_dB=interp1(LowerSidelobeGoal_
u+[0:size(LowerSidelobeGoal_u,2)-1]*eps, LowerSidelobeGoal_
dB, SineTheta);
UpperEnvelope_dB=interp1(UpperSidelobeGoal_
u+[0:size(UpperSidelobeGoal_u,2)-1]*eps, UpperSidelobeGoal_
dB, SineTheta);
LowerEnvelope=10.^(LowerEnvelope_dB/20)';
UpperEnvelope=10.^(UpperEnvelope_dB/20)';
ElementWgtPower=CosineTheta.^CosineElementFactorExponent;
ElementWgtVoltage=sqrt(ElementWgtPower)';
tic;

%%%%%%%%%%%%%%%%%%%%%%%%%%%%%%%%%%%%%%%%
% Initialize Population
%%%%%%%%%%%%%%%%%%%%%%%%%%%%%%%%%%%%%%%%
CurrentGeneration=rand(Norder,Npopulation);
%%%%%%%%%%%%%%%%%%%%%%%%%%%%%%%%%%%%%%%%
% Score Population
%%%%%%%%%%%%%%%%%%%%%%%%%%%%%%%%%%%%%%%%
```

```matlab
for kk=1:Npopulation
  MagWgts=CurrentGeneration(1:Nelements,kk).*MagRange+Mag
  Min;
  PhsWgts=CurrentGeneration(Nelements+1:2*Nelements,kk).*Phs
  Range+PhsMin;
  ComplexWgts=(MagWgts.*exp(i*PhsWgts));
  CurrentGenerationScores(kk)=simpleCostFunction(...
  ComplexWgts,ElementWgtVoltage,LowerEnvelope,UpperEnvel
  ope,...
  AutomaticallyExemptMainBeam,EnforceSymmetry);
end
%%%%%%%%%%%%%%%%%%%%%%%%%%%%%%%%%%%%%
% Get Ready to Optimize
%%%%%%%%%%%%%%%%%%%%%%%%%%%%%%%%%%%%%
[SortFitness,SortIndx]=sort(CurrentGenerationScores);
CurrentGeneration=CurrentGeneration(:,SortIndx);
CurrentGenerationScores=CurrentGenerationScores(SortIndx);
GlobalBestScore=CurrentGenerationScores(1);


%%%%%%%%%%%%%%%%%%%%%%%%%%%%%%%%%%%%%%%
% Main Optimization Loop
%%%%%%%%%%%%%%%%%%%%%%%%%%%%%%%%%%%%%%%
for ii=1:Niterations
  t=(ii-1)/(Niterations-1);
  MutationRange=MutationRangeMax*(exp(-
  MutationRangeDecay*t)-exp(-MutationRangeDecay));
  %%%%%%%%%%%%%%%%%%%%%%%%%%%%%%%%%%%%%%%%%
  % For Each Marriage:
  %%%%%%%%%%%%%%%%%%%%%%%%%%%%%%%%%%%%%%%%%
  for kk=1:Nmarriages
    %%%%%%%%%%%%%%%%%%%%%%%%%%%%%%%%%%%%%%%%%%%
    % Choose Mom
    %%%%%%%%%%%%%%%%%%%%%%%%%%%%%%%%%%%%%%%%%%%
    PermutedIndx=randperm(TournamentEligible);
    CompetitorFitness=CurrentGeneration(PermutedIndx(1:
    TournamentCompetitors));
    WinnerIndx=find(CompetitorFitness==min(CompetitorFitn
    ess));
    Mom=CurrentGeneration(:,PermutedIndx(WinnerIndx(1)));
    %%%%%%%%%%%%%%%%%%%%%%%%%%%%%%%%%%%%%%%%%%%
    % Choose Dad
    %%%%%%%%%%%%%%%%%%%%%%%%%%%%%%%%%%%%%%%%%%%
    PermutedIndx=randperm(TournamentEligible);
    CompetitorFitness=CurrentGeneration(PermutedIndx(1:
    TournamentCompetitors));
    WinnerIndx=find(CompetitorFitness==min(CompetitorFitn
    ess));
    Dad=CurrentGeneration(:,PermutedIndx(WinnerIndx(1)));
```

```
%%%%%%%%%%%%%%%%%%%%%%%%%%%%%%%%%%%%%%%%
% Mix Mom and Dad to Make Kids
%%%%%%%%%%%%%%%%%%%%%%%%%%%%%%%%%%%%%%%%
Son=Dad;
Daughter=Mom;
PermutedIndx=randperm(Norder-1)+1;
CrossoverPoints=sort(PermutedIndx(1:Ncrossovers));
for CrossOverCount=1:Ncrossovers
  Transfer=Son(CrossoverPoints(CrossOverCount):Norder);
Son(CrossoverPoints(CrossOverCount):Norder)=Daughter(Crossov
erPoints(CrossOverCount):Norder);
  Daughter(CrossoverPoints(CrossOverCount):Norder)=Trans
  fer;
end
%%%%%%%%%%%%%%%%%%%%%%%%%%%%%%%%%%%%%%%%
% Mutate Son
%%%%%%%%%%%%%%%%%%%%%%%%%%%%%%%%%%%%%%%%
PermutedIndx=randperm(Norder);
MutationDecision=zeros(Norder,1);
MutationDecision(PermutedIndx(1:Nmutations))=1;
MutatedMax=min(Son+MutationRange/2, 1);
MutatedMin=max(Son-MutationRange/2, 0);
Mutations=rand(Norder,1).*(MutatedMax-
MutatedMin)+MutatedMin;
Children(:,2*kk-1)=Son.*(1-MutationDecision)+Mutations.*
MutationDecision;
%%%%%%%%%%%%%%%%%%%%%%%%%%%%%%%%%%%%%%%%
% Mutate Daughter
%%%%%%%%%%%%%%%%%%%%%%%%%%%%%%%%%%%%%%%%
PermutedIndx=randperm(Norder);
MutationDecision=zeros(Norder,1);
MutationDecision(PermutedIndx(1:Nmutations))=1;
MutatedMax=min(Daughter+MutationRange/2, 1);
MutatedMin=max(Daughter-MutationRange/2, 0);
Mutations=rand(Norder,1).*(MutatedMax-
MutatedMin)+MutatedMin;
Children(:,2*kk)=Daughter.*(1-MutationDecision)+Mutation
s.*MutationDecision;
end
%%%%%%%%%%%%%%%%%%%%%%%%%%%%%%%%%%%%%%%%
% Score Children
%%%%%%%%%%%%%%%%%%%%%%%%%%%%%%%%%%%%%%%%
for kk=1:2*Nmarriages
  MagWgts=Children(1:Nelements,kk).*MagRange+MagMin;
  PhsWgts=Children(Nelements+1:2*Nelements,kk).*PhsRange+P
  hsMin;
  ComplexWgts=(MagWgts.*exp(i*PhsWgts));
  ChildrenScores(kk)=simpleCostFunction(...
```

```
      ComplexWgts,ElementWgtVoltage,LowerEnvelope,UpperEnvel
      ope,...
      AutomaticallyExemptMainBeam,EnforceSymmetry);
  end
  %%%%%%%%%%%%%%%%%%%%%%%%%%%%%%%%%%%%%%%%
  % Survival of the Fittest
  %%%%%%%%%%%%%%%%%%%%%%%%%%%%%%%%%%%%%%%%
  CombinedGenerations=[CurrentGeneration Children];
  CombinedScores=[CurrentGenerationScores ChildrenScores];
  [SortFitness,SortIndx]=sort(CombinedScores);
  SurvivorsIndx=SortIndx(1:Npopulation);
  CurrentGeneration=CombinedGenerations(:,SurvivorsIndx);
  CurrentGenerationScores=CombinedScores(SurvivorsIndx);
  GlobalBestScore=CurrentGenerationScores(1);
  %%%%%%%%%%%%%%%%%%%%%%%%%%%%%%%%%%%%%%%%
  % Save Performance Results
  %%%%%%%%%%%%%%%%%%%%%%%%%%%%%%%%%%%%%%%%
  SavePopulationAverageScore(ii)=mean(CurrentGenerationScores);
  SavePopulationBestScores(ii)=GlobalBestScore;
end
GlobalBestIndividual=CurrentGeneration(:,1);

%%%%%%%%%%%%%%%%%%%%%%%%%%%%%%%%%%%%%%%%%%
% Calculate Best Pattern
%%%%%%%%%%%%%%%%%%%%%%%%%%%%%%%%%%%%%%%%%%
NPatternPointsFine=NPatternPoints*8;
SineThetaFine=[-1:2/NPatternPointsFine:1-2/
NPatternPointsFine];
CosineThetaFine=sqrt(1-SineThetaFine.^2);
ElementWgtPowerFine=CosineThetaFine.^CosineElementFactorExpo
nent;
ElementWgtVoltageFine=sqrt(ElementWgtPowerFine).';
ComplexWgts=(GlobalBestIndividual(1:Nelements).*MagRange+Mag
Min).*...
  exp(i*(GlobalBestIndividual(Nelements+1:2*Nelements).*PhsR
ange+PhsMin));
if EnforceSymmetry
  ComplexWgts(Nelements/2+1:Nelements)=flipud(ComplexWgts(1:
  Nelements/2));
end
ComplexWgts=ComplexWgts/max(abs(ComplexWgts));
BestPattern=fftshift (fft(ComplexWgts,NPatternPointsFine)).*
ElementWgtVoltageFine;
BestPattern_dB=20*log10(abs(BestPattern)+eps);
BestPatternNorm_dB=BestPattern_dB-max(BestPattern_dB);
ElapsedTimeMinutes=toc/60;

%%%%%%%%%%%%%%%%%%%%%%%%%%%%%%%%%%%%%%%%%%
% Plot Results
```

```
%%%%%%%%%%%%%%%%%%%%%%%%%%%%%%%%%%%%%%%%%%
figure
figscale=70; figoffsetx=20; figoffsety=20;
set(gcf,'Position',[figoffsetx figoffsety round(11.25*figsca
le+figoffsetx) round(6.75*figscale+figoffsety)])
fontsize=12;
subplot(2,2,1)
set(gca,'FontSize',fontsize)
plot([1:Niterations],SavePopulationBestScores,[1:Niterations
],SavePopulationAverageScore,'LineWidth',2)
xlabel('Iteration #')
ylabel('Cost Function (arbitrary units)')
legend('Best','Average')
axis tight
subplot(2,2,2)
[ax,h1,h2]=plotyy([1:Nelements],abs(ComplexWgts),[1:Nelement
s],angle(ComplexWgts)-min(angle(ComplexWgts)));
set(h1,'Marker','.','LineStyle','none');
set(h2,'Marker','.','LineStyle','none');
axes(ax(1));
set(gca,'FontSize',fontsize)
ylabel('Amplitude (Volts) (blue)')
ylim([0 1])
axes(ax(2));
set(gca,'FontSize',fontsize)
ylabel('Phase (radians) (green)')
title('Aperture weights from Genetic Algorithm')
subplot(2,1,2)
set(gca,'FontSize',fontsize)
plot(SineThetaFine,BestPatternNorm_dB,'LineWidth',2);
hold on
plot(LowerSidelobeGoal_u, LowerSidelobeGoal_dB,
'r-.','LineWidth',2)
plot(UpperSidelobeGoal_u, UpperSidelobeGoal_dB,
'r:','LineWidth',2)
axis tight
ylim([-80 0])
grid on
title(['Far field Pattern (Target sidelobe constraints in
red). Elapsed time = ' num2str(ElapsedTimeMinutes) '
minutes.'])
ylabel('Magnitude (dB)')
xlabel('Sin(\theta)')
```

4.4.4   *simpleGA_PhaseOnlyEx.m*

```
clc
close all
clear all
```

```
%%%%%%%%%%%%%%%%%%%%%%%%%%%%%%%%%%%%%%%%
% Antenna Description
%%%%%%%%%%%%%%%%%%%%%%%%%%%%%%%%%%%%%%%%
Nelements=100;
CosineElementFactorExponent=1.2;
%%%%%%%%%%%%%%%%%%%%%%%%%%%%%%%%%%%%%%%%
% Excitation Constraints
%%%%%%%%%%%%%%%%%%%%%%%%%%%%%%%%%%%%%%%%
% Taylor.m is included with the code in Chapters 1, 2, and 3
MagMin(1:Nelements,1)=Taylor(Nelements,30,5);
MagMax(1:Nelements,1)=Taylor(Nelements,30,5);
PhsMin(1:Nelements,1)=0;
PhsMax(1:Nelements,1)=pi/2;
EnforceSymmetry=0; % 1 for symmetric distrib, 0 for asymmetric
%%%%%%%%%%%%%%%%%%%%%%%%%%%%%%%%%%%%%%%%
% Pattern Goals
%%%%%%%%%%%%%%%%%%%%%%%%%%%%%%%%%%%%%%%%
UpperSidelobeGoal_u= [ -1 -.03 .03 .4 .4 .5 .5 1 ];
UpperSidelobeGoal_dB=[-55 -30 -30 -39.5 -60 -60 -42.1 -55];
LowerSidelobeGoal_u= [ -1 1];
LowerSidelobeGoal_dB=[-80 -80];
AutomaticallyExemptMainBeam=1; % 1 to exempt main beam
from goals
NPatternPoints=512;
%%%%%%%%%%%%%%%%%%%%%%%%%%%%%%%%%%%%%%%%
% Optimization Parameters
%%%%%%%%%%%%%%%%%%%%%%%%%%%%%%%%%%%%%%%%
Niterations=1000;
Npopulation=50;
Nmarriages=25;
TournamentEligible=10;
TournamentCompetitors=5;
Ncrossovers=2;
MutationProbability=0.04;
MutationRangeMax=0.2;
MutationRangeDecay=1.5;
Norder=2*Nelements;
Nmutations=round(MutationProbability*Norder);
%%%%%%%%%%%%%%%%%%%%%%%%%%%%%%%%%%%%%%%%
% Miscellaneous Initialization
%%%%%%%%%%%%%%%%%%%%%%%%%%%%%%%%%%%%%%%%
MagRange=MagMax-MagMin;
PhsRange=PhsMax-PhsMin;
SineTheta=[-1:2/NPatternPoints:1-2/NPatternPoints];
CosineTheta=sqrt(1-SineTheta.^2);
LowerEnvelope_dB=interp1(LowerSidelobeGoal_
u+[0:size(LowerSidelobeGoal_u,2)-1]*eps, LowerSidelobeGoal_
dB, SineTheta);
```

```
UpperEnvelope_dB=interp1(UpperSidelobeGoal_
u+[0:size(UpperSidelobeGoal_u,2)-1]*eps, UpperSidelobeGoal_
dB, SineTheta);
LowerEnvelope=10.^(LowerEnvelope_dB/20)';
UpperEnvelope=10.^(UpperEnvelope_dB/20)';
ElementWgtPower=CosineTheta.^CosineElementFactorExponent;
ElementWgtVoltage=sqrt(ElementWgtPower)';
tic;

%%%%%%%%%%%%%%%%%%%%%%%%%%%%%%%%%%%%%
% Initialize Population
%%%%%%%%%%%%%%%%%%%%%%%%%%%%%%%%%%%%%
CurrentGeneration=rand(Norder,Npopulation);
%%%%%%%%%%%%%%%%%%%%%%%%%%%%%%%%%%%%%
% Score Population
%%%%%%%%%%%%%%%%%%%%%%%%%%%%%%%%%%%%%
for kk=1:Npopulation
  MagWgts=CurrentGeneration(1:Nelements,kk).*MagRange+Mag
  Min;
  PhsWgts=CurrentGeneration(Nelements+1:2*Nelements,kk).*Phs
  Range+PhsMin;
  ComplexWgts=(MagWgts.*exp(i*PhsWgts));
  CurrentGenerationScores(kk)=simpleCostFunction(...
    ComplexWgts,ElementWgtVoltage,LowerEnvelope,UpperEnvel
    ope,...
    AutomaticallyExemptMainBeam,EnforceSymmetry);
end
%%%%%%%%%%%%%%%%%%%%%%%%%%%%%%%%%%%%%
% Get Ready to Optimize
%%%%%%%%%%%%%%%%%%%%%%%%%%%%%%%%%%%%%
[SortFitness,SortIndx]=sort(CurrentGenerationScores);
CurrentGeneration=CurrentGeneration(:,SortIndx);
CurrentGenerationScores=CurrentGenerationScores(SortIndx);
GlobalBestScore=CurrentGenerationScores(1);

%%%%%%%%%%%%%%%%%%%%%%%%%%%%%%%%%%%%%
% Main Optimization Loop
%%%%%%%%%%%%%%%%%%%%%%%%%%%%%%%%%%%%%
for ii=1:Niterations
  t=(ii-1)/(Niterations-1);
  MutationRange=MutationRangeMax*(exp(-
  MutationRangeDecay*t)-exp(-MutationRangeDecay));
  %%%%%%%%%%%%%%%%%%%%%%%%%%%%%%%%%%%
  % For Each Marriage:
  %%%%%%%%%%%%%%%%%%%%%%%%%%%%%%%%%%%
  for kk=1:Nmarriages
    %%%%%%%%%%%%%%%%%%%%%%%%%%%%%%%%%%%
    % Choose Mom
```

```
%%%%%%%%%%%%%%%%%%%%%%%%%%%%%%%%%%%%%
PermutedIndx=randperm(TournamentEligible);
CompetitorFitness=CurrentGeneration(PermutedIndx(1:
TournamentCompetitors));
WinnerIndx=find(CompetitorFitness==min(CompetitorFitn
ess));
Mom=CurrentGeneration(:,PermutedIndx(WinnerIndx(1)));
%%%%%%%%%%%%%%%%%%%%%%%%%%%%%%%%%%%%%
% Choose Dad
%%%%%%%%%%%%%%%%%%%%%%%%%%%%%%%%%%%%%
PermutedIndx=randperm(TournamentEligible);
CompetitorFitness=CurrentGeneration(PermutedIndx(1:
TournamentCompetitors));
WinnerIndx=find(CompetitorFitness==min(CompetitorFitn
ess));
Dad=CurrentGeneration(:,PermutedIndx(WinnerIndx(1)));
%%%%%%%%%%%%%%%%%%%%%%%%%%%%%%%%%%%%%
% Mix Mom and Dad to Make Kids
%%%%%%%%%%%%%%%%%%%%%%%%%%%%%%%%%%%%%
Son=Dad;
Daughter=Mom;
PermutedIndx=randperm(Norder-1)+1;
CrossoverPoints=sort(PermutedIndx(1:Ncrossovers));
for CrossOverCount=1:Ncrossovers
  Transfer=Son(CrossoverPoints(CrossOverCount):Norder);
Son(CrossoverPoints(CrossOverCount):Norder)=Daughter(Crossov
erPoints(CrossOverCount):Norder);
  Daughter(CrossoverPoints(CrossOverCount):Norder)=Trans
  fer;
end
%%%%%%%%%%%%%%%%%%%%%%%%%%%%%%%%%%%%%
% Mutate Son
%%%%%%%%%%%%%%%%%%%%%%%%%%%%%%%%%%%%%
PermutedIndx=randperm(Norder);
MutationDecision=zeros(Norder,1);
MutationDecision(PermutedIndx(1:Nmutations))=1;
MutatedMax=min(Son+MutationRange/2, 1);
MutatedMin=max(Son-MutationRange/2, 0);
Mutations=rand(Norder,1).*(MutatedMax-
MutatedMin)+MutatedMin;
Children(:,2*kk-1)=Son.*(1-MutationDecision)+Mutations.*
MutationDecision;
%%%%%%%%%%%%%%%%%%%%%%%%%%%%%%%%%%%%%
% Mutate Daughter
%%%%%%%%%%%%%%%%%%%%%%%%%%%%%%%%%%%%%
PermutedIndx=randperm(Norder);
MutationDecision=zeros(Norder,1);
MutationDecision(PermutedIndx(1:Nmutations))=1;
```

```
    MutatedMax=min(Daughter+MutationRange/2, 1);
    MutatedMin=max(Daughter-MutationRange/2, 0);
    Mutations=rand(Norder,1).*(MutatedMax-
    MutatedMin)+MutatedMin;
    Children(:,2*kk)=Daughter.*(1-MutationDecision)+Mutation
    s.*MutationDecision;
  end
  %%%%%%%%%%%%%%%%%%%%%%%%%%%%%%%%%%%%%%
  % Score Children
  %%%%%%%%%%%%%%%%%%%%%%%%%%%%%%%%%%%%%%
  for kk=1:2*Nmarriages
    MagWgts=Children(1:Nelements,kk).*MagRange+MagMin;
    PhsWgts=Children(Nelements+1:2*Nelements,kk).*PhsRange+P
    hsMin;
    ComplexWgts=(MagWgts.*exp(i*PhsWgts));
    ChildrenScores(kk)=simpleCostFunction(...
      ComplexWgts,ElementWgtVoltage,LowerEnvelope,UpperEnvel
      ope,...
      AutomaticallyExemptMainBeam,EnforceSymmetry);
  end
  %%%%%%%%%%%%%%%%%%%%%%%%%%%%%%%%%%%%%%
  % Survival of the Fittest
  %%%%%%%%%%%%%%%%%%%%%%%%%%%%%%%%%%%%%%
  CombinedGenerations=[CurrentGeneration Children];
  CombinedScores=[CurrentGenerationScores ChildrenScores];
  [SortFitness,SortIndx]=sort(CombinedScores);
  SurvivorsIndx=SortIndx(1:Npopulation);
  CurrentGeneration=CombinedGenerations(:,SurvivorsIndx);
  CurrentGenerationScores=CombinedScores(SurvivorsIndx);
  GlobalBestScore=CurrentGenerationScores(1);
  %%%%%%%%%%%%%%%%%%%%%%%%%%%%%%%%%%%%%%
  % Save Performance Results
  %%%%%%%%%%%%%%%%%%%%%%%%%%%%%%%%%%%%%%
  SavePopulationAverageScore(ii)=mean(CurrentGenerationSco
  res);
  SavePopulationBestScores(ii)=GlobalBestScore;
end
GlobalBestIndividual=CurrentGeneration(:,1);

%%%%%%%%%%%%%%%%%%%%%%%%%%%%%%%%%%%%%%%%
% Calculate Best Pattern
%%%%%%%%%%%%%%%%%%%%%%%%%%%%%%%%%%%%%%%%
NPatternPointsFine=NPatternPoints*8;
SineThetaFine=[-1:2/NPatternPointsFine:1-2/
NPatternPointsFine];
CosineThetaFine=sqrt(1-SineThetaFine.^2);
ElementWgtPowerFine=CosineThetaFine.^CosineElementFactorExpo
nent;
```

```
ElementWgtVoltageFine=sqrt(ElementWgtPowerFine).';
ComplexWgts=(GlobalBestIndividual(1:Nelements).*MagRange+Mag
Min).*...
  exp(i*(GlobalBestIndividual(Nelements+1:2*Nelements).*PhsR
  ange+PhsMin));
if EnforceSymmetry
  ComplexWgts(Nelements/2+1:Nelements)=flipud(ComplexWgts(1:
  Nelements/2));
end
ComplexWgts=ComplexWgts/max(abs(ComplexWgts));
BestPattern=fftshift (fft(ComplexWgts,NPatternPointsFine)).*
ElementWgtVoltageFine;
BestPattern_dB=20*log10(abs(BestPattern)+eps);
BestPatternNorm_dB=BestPattern_dB-max(BestPattern_dB);
ElapsedTimeMinutes=toc/60;

%%%%%%%%%%%%%%%%%%%%%%%%%%%%%%%%%%%%%%%%
% Plot Results
%%%%%%%%%%%%%%%%%%%%%%%%%%%%%%%%%%%%%%%%
figure
figscale=70; figoffsetx=20; figoffsety=20;
set(gcf,'Position',[figoffsetx figoffsety round(11.25*figsca
le+figoffsetx) round(6.75*figscale+figoffsety)])
fontsize=12;
subplot(2,2,1)
set(gca,'FontSize',fontsize)
plot([1:Niterations],SavePopulationBestScores,[1:Niterations
],SavePopulationAverageScore,'LineWidth',2)
xlabel('Iteration #')
ylabel('Cost Function (arbitrary units)')
legend('Best','Average')
axis tight
subplot(2,2,2)
[ax,h1,h2]=plotyy([1:Nelements],abs(ComplexWgts),[1:Nelement
s],angle(ComplexWgts)-min(angle(ComplexWgts)));
set(h1,'Marker','.','LineStyle','none');
set(h2,'Marker','.','LineStyle','none');
axes(ax(1));
set(gca,'FontSize',fontsize)
ylabel('Amplitude (Volts) (blue)')
ylim([0 1])
axes(ax(2));
set(gca,'FontSize',fontsize)
ylabel('Phase (radians) (green)')
title('Aperture weights from Genetic Algorithm')
subplot(2,1,2)
set(gca,'FontSize',fontsize)
plot(SineThetaFine,BestPatternNorm_dB,'LineWidth',2);
hold on
```

```
plot(LowerSidelobeGoal_u, LowerSidelobeGoal_dB,
'r-.','LineWidth',2)
plot(UpperSidelobeGoal_u, UpperSidelobeGoal_dB,
'r:','LineWidth',2)
axis tight
ylim([-80 0])
grid on
title(['Far field Pattern (Target sidelobe constraints in
red). Elapsed time = ' num2str(ElapsedTimeMinutes) '
minutes.'])
ylabel('Magnitude (dB)')
xlabel('Sin(\theta)')
```

4.4.5   *simplePS_AmpTaperEx.m*

```
clc
close all
clear all
%%%%%%%%%%%%%%%%%%%%%%%%%%%%%%%%%%%%%%
% Antenna Description
%%%%%%%%%%%%%%%%%%%%%%%%%%%%%%%%%%%%%%
Nelements=100;
CosineElementFactorExponent=1.2;
%%%%%%%%%%%%%%%%%%%%%%%%%%%%%%%%%%%%%%
% Excitation Constraints
%%%%%%%%%%%%%%%%%%%%%%%%%%%%%%%%%%%%%%
MagMin(1:Nelements,1)=0;
MagMax(1:Nelements,1)=1;
PhsMin(1:Nelements,1)=0;
PhsMax(1:Nelements,1)=0;
EnforceSymmetry=1; % 1 for symmetric distrib, 0 for asymmetric
%%%%%%%%%%%%%%%%%%%%%%%%%%%%%%%%%%%%%%
% Pattern Goals
%%%%%%%%%%%%%%%%%%%%%%%%%%%%%%%%%%%%%%
UpperSidelobeGoal_u= [ -1 -.03 .03 .4 .4 .5 .5 1 ];
UpperSidelobeGoal_dB=[-55 -30 -30 -39.5 -60 -60 -42.1 -55];
LowerSidelobeGoal_u= [ -1 1];
LowerSidelobeGoal_dB=[-80 -80];
AutomaticallyExemptMainBeam=1; % 1 to exempt main beam
from goals
NPatternPoints=512;
%%%%%%%%%%%%%%%%%%%%%%%%%%%%%%%%%%%%%%
% Optimization Parameters
%%%%%%%%%%%%%%%%%%%%%%%%%%%%%%%%%%%%%%
Niterations=1000;
Npopulation=50;
Phi1=2;
```

```
Phi2=2;
W=.4;
VRMSmax=.3;
Norder=2*Nelements;
%%%%%%%%%%%%%%%%%%%%%%%%%%%%%%%%%%%%%%
% Miscellaneous Initialization
%%%%%%%%%%%%%%%%%%%%%%%%%%%%%%%%%%%%%%
MagRange=MagMax-MagMin;
PhsRange=PhsMax-PhsMin;
SineTheta=[-1:2/NPatternPoints:1-2/NPatternPoints];
CosineTheta=sqrt(1-SineTheta.^2);
LowerEnvelope_dB=interp1(LowerSidelobeGoal_
u+[0:size(LowerSidelobeGoal_u,2)-1]*eps, LowerSidelobeGoal_
dB, SineTheta);
UpperEnvelope_dB=interp1(UpperSidelobeGoal_
u+[0:size(UpperSidelobeGoal_u,2)-1]*eps, UpperSidelobeGoal_
dB, SineTheta);
LowerEnvelope=10.^(LowerEnvelope_dB/20)';
UpperEnvelope=10.^(UpperEnvelope_dB/20)';
ElementWgtPower=CosineTheta.^CosineElementFactorExponent;
ElementWgtVoltage=sqrt(ElementWgtPower)';
tic;


%%%%%%%%%%%%%%%%%%%%%%%%%%%%%%%%%%%%%%
% Initialize Particle Swarm
%%%%%%%%%%%%%%%%%%%%%%%%%%%%%%%%%%%%%%
Swarm=rand(Norder,Npopulation);
SwarmVelocity=rand(Norder,Npopulation)*2-1;
%%%%%%%%%%%%%%%%%%%%%%%%%%%%%%%%%%%%%%
% Score Particles
%%%%%%%%%%%%%%%%%%%%%%%%%%%%%%%%%%%%%%
for kk=1:Npopulation
  MagWgts=Swarm(1:Nelements,kk).*MagRange+MagMin;
  PhsWgts=Swarm(Nelements+1:2*Nelements,kk).*PhsRange+Phs
  Min;
  ComplexWgts=(MagWgts.*exp(i*PhsWgts));
  SwarmScores(kk)=simpleCostFunction(...
    ComplexWgts,ElementWgtVoltage,LowerEnvelope,UpperEnvel
    ope,...
    AutomaticallyExemptMainBeam,EnforceSymmetry);
end
%%%%%%%%%%%%%%%%%%%%%%%%%%%%%%%%%%%%%%
% Get Ready to Optimize
%%%%%%%%%%%%%%%%%%%%%%%%%%%%%%%%%%%%%%
SwarmScoreMemory=SwarmScores;
SwarmLocalBest=Swarm;
GlobalBestScore=min(SwarmScores);
IndexOfBest=find(SwarmScores==GlobalBestScore);
SwarmGlobalBest=Swarm(:,IndexOfBest(1));
```

```
%%%%%%%%%%%%%%%%%%%%%%%%%%%%%%%%%%%%%%%%
% Main Optimization Loop
%%%%%%%%%%%%%%%%%%%%%%%%%%%%%%%%%%%%%%%%
for ii=1:Niterations
  %%%%%%%%%%%%%%%%%%%%%%%%%%%%%%%%%%%%%%%%%%
  % For Each Particle:
  %%%%%%%%%%%%%%%%%%%%%%%%%%%%%%%%%%%%%%%%%%
  for kk=1:Npopulation
    %%%%%%%%%%%%%%%%%%%%%%%%%%%%%%%%%%%%%%%%%%%%
    % Update Velocity
    %%%%%%%%%%%%%%%%%%%%%%%%%%%%%%%%%%%%%%%%%%%%
    SwarmSelfKnowledge=Phi1*rand(Norder,1).*...
      (SwarmLocalBest(:,kk)-Swarm(:,kk));
    SwarmSocialKnowledge=Phi2*rand(Norder,1).*...
      (SwarmGlobalBest-Swarm(:,kk));
    SwarmVelocity(:,kk)=W*SwarmVelocity(:,kk)+...
      SwarmSelfKnowledge+SwarmSocialKnowledge;
    VRMS=sqrt(sum(SwarmVelocity(:,kk).^2));
    if VRMS>VRMSmax
      SwarmVelocity(:,kk)=SwarmVelocity(:,kk)/VRMS*VRMSmax;
    end
    %%%%%%%%%%%%%%%%%%%%%%%%%%%%%%%%%%%%%%%%%%%%
    % Update Position
    %%%%%%%%%%%%%%%%%%%%%%%%%%%%%%%%%%%%%%%%%%%%
    Swarm(:,kk)=Swarm(:,kk)+SwarmVelocity(:,kk);
    Swarm(:,kk)=max(min(Swarm(:,kk),1),0);
    %%%%%%%%%%%%%%%%%%%%%%%%%%%%%%%%%%%%%%%%%%%%
    % Update Score
    %%%%%%%%%%%%%%%%%%%%%%%%%%%%%%%%%%%%%%%%%%%%
    MagWgts=Swarm(1:Nelements,kk).*MagRange+MagMin;
    PhsWgts=Swarm(Nelements+1:2*Nelements,kk).*PhsRange+Phs
    Min;
    ComplexWgts=(MagWgts.*exp(i*PhsWgts));
    SwarmScores(kk)=simpleCostFunction(...
      ComplexWgts,ElementWgtVoltage,LowerEnvelope,UpperEnvel
      ope,...
      AutomaticallyExemptMainBeam,EnforceSymmetry);
    %%%%%%%%%%%%%%%%%%%%%%%%%%%%%%%%%%%%%%%%%%%%
    % Update Particle Memory
    %%%%%%%%%%%%%%%%%%%%%%%%%%%%%%%%%%%%%%%%%%%%
    if SwarmScores(kk)<SwarmScoreMemory(kk)
      SwarmScoreMemory(kk)=SwarmScores(kk);
      SwarmLocalBest(:,kk)=Swarm(:,kk);
    end
    %%%%%%%%%%%%%%%%%%%%%%%%%%%%%%%%%%%%%%%%%%%%
    % Update Group Knowledge
    %%%%%%%%%%%%%%%%%%%%%%%%%%%%%%%%%%%%%%%%%%%%
    if SwarmScores(kk)<GlobalBestScore
```

```
    SwarmGlobalBest=Swarm(:,kk);
    GlobalBestScore=SwarmScores(kk);
  end
 end
 %%%%%%%%%%%%%%%%%%%%%%%%%%%%%%%%%%%%%
 % Save Performance Results
 %%%%%%%%%%%%%%%%%%%%%%%%%%%%%%%%%%%%%
 SaveSwarmAverageScore(ii)=mean(SwarmScores);
 SaveSwarmBestScores(ii)=GlobalBestScore;
end

%%%%%%%%%%%%%%%%%%%%%%%%%%%%%%%%%%%%%%
% Calculate Best Pattern
%%%%%%%%%%%%%%%%%%%%%%%%%%%%%%%%%%%%%%
NPatternPointsFine=NPatternPoints*8;
SineThetaFine=[-1:2/NPatternPointsFine:1-2/
NPatternPointsFine];
CosineThetaFine=sqrt(1-SineThetaFine.^2);
ElementWgtPowerFine=CosineThetaFine.^CosineElementFactorExpo
nent;
ElementWgtVoltageFine=sqrt(ElementWgtPowerFine).';
ComplexWgts=(SwarmGlobalBest(1:Nelements).*MagRange+Mag
Min).*...
  exp(i*(SwarmGlobalBest(Nelements+1:2*Nelements).*PhsRange+
  PhsMin));
if EnforceSymmetry
  ComplexWgts(Nelements/2+1:Nelements)=flipud(ComplexWgts(1:
  Nelements/2));
end
ComplexWgts=ComplexWgts/max(abs(ComplexWgts));
BestPattern=fftshift (ff t(ComplexWgts,NPatternPointsFine)).
*ElementWgtVoltageFine;
BestPattern_dB=20*log10(abs(BestPattern)+eps);
BestPatternNorm_dB=BestPattern_dB-max(BestPattern_dB);
ElapsedTimeMinutes=toc/60;

%%%%%%%%%%%%%%%%%%%%%%%%%%%%%%%%%%%%%%%%
% Plot Results
%%%%%%%%%%%%%%%%%%%%%%%%%%%%%%%%%%%%%%%%
figure
figscale=70; figoffsetx=20; figoffsety=20;
set(gcf,'Position',[figoffsetx figoffsety round(11.25*figsca
le+figoffsetx) round(6.75*figscale+figoffsety)])
fontsize=12;
subplot(2,2,1)
set(gca,'FontSize',fontsize)
plot([1:Niterations],SaveSwarmBestScores,[1:Niterations],Sav
eSwarmAverageScore,'LineWidth',2)
xlabel('Iteration #')
```

```
ylabel('Cost Function (arbitrary units)')
legend('Best','Average')
axis tight
subplot(2,2,2)
[ax,h1,h2]=plotyy([1:Nelements],abs(ComplexWgts),[1:Nelement
s],angle(ComplexWgts)-min(angle(ComplexWgts)));
set(h1,'Marker','.','LineStyle','none');
set(h2,'Marker','.','LineStyle','none');
axes(ax(1));
set(gca,'FontSize',fontsize)
ylabel('Amplitude (Volts) (blue)')
ylim([0 1])
axes(ax(2));
set(gca,'FontSize',fontsize)
ylabel('Phase (radians) (green)')
title('Aperture weights from Particle Swarm')
subplot(2,1,2)
set(gca,'FontSize',fontsize)
plot(SineThetaFine,BestPatternNorm_dB,'LineWidth',2);
hold on
plot(LowerSidelobeGoal_u, LowerSidelobeGoal_dB,
'r-.','LineWidth',2)
plot(UpperSidelobeGoal_u, UpperSidelobeGoal_dB,
'r:','LineWidth',2)
axis tight
ylim([-80 0])
grid on
title(['Far field Pattern (Target sidelobe constraints in
red). Elapsed time = ' num2str(ElapsedTimeMinutes) '
minutes.'])
ylabel('Magnitude (dB)')
xlabel('Sin(\theta)')
```

4.4.6   *simplePS_FlattopEx.m*

```
clc
close all
clear all
%%%%%%%%%%%%%%%%%%%%%%%%%%%%%%%%%%%%%%
% Antenna Description
%%%%%%%%%%%%%%%%%%%%%%%%%%%%%%%%%%%%%%
Nelements=100;
CosineElementFactorExponent=1.2;
%%%%%%%%%%%%%%%%%%%%%%%%%%%%%%%%%%%%%%
% Excitation Constraints
%%%%%%%%%%%%%%%%%%%%%%%%%%%%%%%%%%%%%%
MagMin(1:Nelements,1)=0;
```

```
MagMax(1:Nelements,1)=1;
PhsMin(1:Nelements,1)=0;
PhsMax(1:Nelements,1)=pi;
EnforceSymmetry=1; % 1 for symmetric distrib, 0 for asymmetric
%%%%%%%%%%%%%%%%%%%%%%%%%%%%%%%%%%%%%%%%%
% Pattern Goals
%%%%%%%%%%%%%%%%%%%%%%%%%%%%%%%%%%%%%%%%%
UpperSidelobeGoal_u= [-1 -.065 -.045 .045 .065 1 ];
UpperSidelobeGoal_dB=[-50 -35 0 0 -35 -50];
LowerSidelobeGoal_u= [ -1 -.036 -.036 .036 .036 1];
LowerSidelobeGoal_dB=[-80 -75 -1 -1 -75 -80];
AutomaticallyExemptMainBeam=0; % 1 to exempt main beam
from goals
NPatternPoints=512;
%%%%%%%%%%%%%%%%%%%%%%%%%%%%%%%%%%%%%%%%%
% Optimization Parameters
%%%%%%%%%%%%%%%%%%%%%%%%%%%%%%%%%%%%%%%%%
Niterations=1000*10;
Npopulation=50;
Phi1=2;
Phi2=2;
W=.4;
VRMSmax=.3;
Norder=2*Nelements;
%%%%%%%%%%%%%%%%%%%%%%%%%%%%%%%%%%%%%%%%%
% Miscellaneous Initialization
%%%%%%%%%%%%%%%%%%%%%%%%%%%%%%%%%%%%%%%%%
MagRange=MagMax-MagMin;
PhsRange=PhsMax-PhsMin;
SineTheta=[-1:2/NPatternPoints:1-2/NPatternPoints];
CosineTheta=sqrt(1-SineTheta.^2);
LowerEnvelope_dB=interp1(LowerSidelobeGoal_
u+[0:size(LowerSidelobeGoal_u,2)-1]*eps, LowerSidelobeGoal_
dB, SineTheta);
UpperEnvelope_dB=interp1(UpperSidelobeGoal_
u+[0:size(UpperSidelobeGoal_u,2)-1]*eps, UpperSidelobeGoal_
dB, SineTheta);
LowerEnvelope=10.^(LowerEnvelope_dB/20)';
UpperEnvelope=10.^(UpperEnvelope_dB/20)';
ElementWgtPower=CosineTheta.^CosineElementFactorExponent;
ElementWgtVoltage=sqrt(ElementWgtPower)';
tic;


%%%%%%%%%%%%%%%%%%%%%%%%%%%%%%%%%%%%%%%%%
% Initialize Particle Swarm
%%%%%%%%%%%%%%%%%%%%%%%%%%%%%%%%%%%%%%%%%
Swarm=rand(Norder,Npopulation);
SwarmVelocity=rand(Norder,Npopulation)*2-1;
```

```
%%%%%%%%%%%%%%%%%%%%%%%%%%%%%%%%%%%%%%%%
% Score Particles
%%%%%%%%%%%%%%%%%%%%%%%%%%%%%%%%%%%%%%%%
for kk=1:Npopulation
  MagWgts=Swarm(1:Nelements,kk).*MagRange+MagMin;
  PhsWgts=Swarm(Nelements+1:2*Nelements,kk).*PhsRange+Phs
  Min;
  ComplexWgts=(MagWgts.*exp(i*PhsWgts));
  SwarmScores(kk)=simpleCostFunction(...
    ComplexWgts,ElementWgtVoltage,LowerEnvelope,UpperEnvel
    ope,...
    AutomaticallyExemptMainBeam,EnforceSymmetry);
end
%%%%%%%%%%%%%%%%%%%%%%%%%%%%%%%%%%%%%%%%
% Get Ready to Optimize
%%%%%%%%%%%%%%%%%%%%%%%%%%%%%%%%%%%%%%%%
SwarmScoreMemory=SwarmScores;
SwarmLocalBest=Swarm;
GlobalBestScore=min(SwarmScores);
IndexOfBest=find(SwarmScores==GlobalBestScore);
SwarmGlobalBest=Swarm(:,IndexOfBest(1));

%%%%%%%%%%%%%%%%%%%%%%%%%%%%%%%%%%%%%%%%
% Main Optimization Loop
%%%%%%%%%%%%%%%%%%%%%%%%%%%%%%%%%%%%%%%%
for ii=1:Niterations
  %%%%%%%%%%%%%%%%%%%%%%%%%%%%%%%%%%%%%%%%
  % For Each Particle:
  %%%%%%%%%%%%%%%%%%%%%%%%%%%%%%%%%%%%%%%%
  for kk=1:Npopulation
  %%%%%%%%%%%%%%%%%%%%%%%%%%%%%%%%%%%%%%%%
  % Update Velocity
  %%%%%%%%%%%%%%%%%%%%%%%%%%%%%%%%%%%%%%%%
  SwarmSelfKnowledge=Phi1*rand(Norder,1).*...
    (SwarmLocalBest(:,kk)-Swarm(:,kk));
  SwarmSocialKnowledge=Phi2*rand(Norder,1).*...
    (SwarmGlobalBest-Swarm(:,kk));
  SwarmVelocity(:,kk)=W*SwarmVelocity(:,kk)+...
    SwarmSelfKnowledge+SwarmSocialKnowledge;
  VRMS=sqrt(sum(SwarmVelocity(:,kk).^2));
  if VRMS>VRMSmax
    SwarmVelocity(:,kk)=SwarmVelocity(:,kk)/VRMS*VRMSmax;
  end
  %%%%%%%%%%%%%%%%%%%%%%%%%%%%%%%%%%%%%%%%
  % Update Position
  %%%%%%%%%%%%%%%%%%%%%%%%%%%%%%%%%%%%%%%%
  Swarm(:,kk)=Swarm(:,kk)+SwarmVelocity(:,kk);
  Swarm(:,kk)=max(min(Swarm(:,kk),1),0);
```

```
%%%%%%%%%%%%%%%%%%%%%%%%%%%%%%%%%%%%%%%
% Update Score
%%%%%%%%%%%%%%%%%%%%%%%%%%%%%%%%%%%%%
MagWgts=Swarm(1:Nelements,kk).*MagRange+MagMin;
PhsWgts=Swarm(Nelements+1:2*Nelements,kk).*PhsRange+Phs
Min;
ComplexWgts=(MagWgts.*exp(i*PhsWgts));
SwarmScores(kk)=simpleCostFunction(...
  ComplexWgts,ElementWgtVoltage,LowerEnvelope,UpperEnvel
  ope,...
  AutomaticallyExemptMainBeam,EnforceSymmetry);
%%%%%%%%%%%%%%%%%%%%%%%%%%%%%%%%%%%%%%%
% Update Particle Memory
%%%%%%%%%%%%%%%%%%%%%%%%%%%%%%%%%%%%%
if SwarmScores(kk)<SwarmScoreMemory(kk)
  SwarmScoreMemory(kk)=SwarmScores(kk);
  SwarmLocalBest(:,kk)=Swarm(:,kk);
end
%%%%%%%%%%%%%%%%%%%%%%%%%%%%%%%%%%%%%%%
% Update Group Knowledge
%%%%%%%%%%%%%%%%%%%%%%%%%%%%%%%%%%%%%
if SwarmScores(kk)<GlobalBestScore
  SwarmGlobalBest=Swarm(:,kk);
  GlobalBestScore=SwarmScores(kk);
end
end
%%%%%%%%%%%%%%%%%%%%%%%%%%%%%%%%%%%%%%%
% Save Performance Results
%%%%%%%%%%%%%%%%%%%%%%%%%%%%%%%%%%%%%
SaveSwarmAverageScore(ii)=mean(SwarmScores);
SaveSwarmBestScores(ii)=GlobalBestScore;
end


%%%%%%%%%%%%%%%%%%%%%%%%%%%%%%%%%%%%%%%%%
% Calculate Best Pattern
%%%%%%%%%%%%%%%%%%%%%%%%%%%%%%%%%%%%%%%%%
NPatternPointsFine=NPatternPoints*8;
SineThetaFine=[-1:2/NPatternPointsFine:1-2/
NPatternPointsFine];
CosineThetaFine=sqrt(1-SineThetaFine.^2);
ElementWgtPowerFine=CosineThetaFine.^CosineElementFactorExpo
nent;
ElementWgtVoltageFine=sqrt(ElementWgtPowerFine).';
ComplexWgts=(SwarmGlobalBest(1:Nelements).*MagRange+Mag
Min).*...
  exp(i*(SwarmGlobalBest(Nelements+1:2*Nelements).*PhsRange+
  PhsMin));
if EnforceSymmetry
```

```
  ComplexWgts(Nelements/2+1:Nelements)=flipud(ComplexWgts(1:
  Nelements/2));
end
ComplexWgts=ComplexWgts/max(abs(ComplexWgts));
BestPattern=fftshift ( fft(ComplexWgts,NPatternPointsFine)).
*ElementWgtVoltageFine;
BestPattern_dB=20*log10(abs(BestPattern)+eps);
BestPatternNorm_dB=BestPattern_dB-max(BestPattern_dB);
ElapsedTimeMinutes=toc/60;

%%%%%%%%%%%%%%%%%%%%%%%%%%%%%%%%%%%%%%
% Plot Results
%%%%%%%%%%%%%%%%%%%%%%%%%%%%%%%%%%%%%%
figure
figscale=70; figoffsetx=20; figoffsety=20;
set(gcf,'Position',[figoffsetx figoffsety round(11.25*figsca
le+figoffsetx) round(6.75*figscale+figoffsety)])
fontsize=12;
subplot(2,2,1)
set(gca,'FontSize',fontsize)
plot([1:Niterations],SaveSwarmBestScores,[1:Niterations],Sav
eSwarmAverageScore,'LineWidth',2)
xlabel('Iteration #')
ylabel('Cost Function (arbitrary units)')
legend('Best','Average')
axis tight
subplot(2,2,2)
[ax,h1,h2]=plotyy([1:Nelements],abs(ComplexWgts),[1:Nelement
s],angle(ComplexWgts)-min(angle(ComplexWgts)));
set(h1,'Marker','.','LineStyle','none');
set(h2,'Marker','.','LineStyle','none');
axes(ax(1));
set(gca,'FontSize',fontsize)
ylabel('Amplitude (Volts) (blue)')
ylim([0 1])
axes(ax(2));
set(gca,'FontSize',fontsize)
ylabel('Phase (radians) (green)')
title('Aperture weights from Particle Swarm')
subplot(2,1,2)
set(gca,'FontSize',fontsize)
plot(SineThetaFine,BestPatternNorm_dB,'LineWidth',2);
hold on
plot(LowerSidelobeGoal_u, LowerSidelobeGoal_dB,
'r-.','LineWidth',2)
plot(UpperSidelobeGoal_u, UpperSidelobeGoal_dB,
'r:','LineWidth',2)
axis tight
```

```
ylim([-80 0])
grid on
title(['Far field Pattern (Target sidelobe constraints in
red). Elapsed time = ' num2str(ElapsedTimeMinutes) '
minutes.'])
ylabel('Magnitude (dB)')
xlabel('Sin(\theta)')
```

4.4.7   *simplePS_PhaseOnlyEx.m*

```
clc
close all
clear all
%%%%%%%%%%%%%%%%%%%%%%%%%%%%%%%%%%%%%%%%
% Antenna Description
%%%%%%%%%%%%%%%%%%%%%%%%%%%%%%%%%%%%%%%%
Nelements=100;
CosineElementFactorExponent=1.2;
%%%%%%%%%%%%%%%%%%%%%%%%%%%%%%%%%%%%%%%%
% Excitation Constraints
%%%%%%%%%%%%%%%%%%%%%%%%%%%%%%%%%%%%%%%%
% Taylor is included with the code in Chapters 1, 2, and 3
MagMin(1:Nelements,1)=Taylor(Nelements,30,5);
MagMax(1:Nelements,1)=Taylor(Nelements,30,5);
PhsMin(1:Nelements,1)=0;
PhsMax(1:Nelements,1)=pi/2;
EnforceSymmetry=0; % 1 for symmetric distrib, 0 for asymmetric
%%%%%%%%%%%%%%%%%%%%%%%%%%%%%%%%%%%%%%%%
% Pattern Goals
%%%%%%%%%%%%%%%%%%%%%%%%%%%%%%%%%%%%%%%%
UpperSidelobeGoal_u= [ -1 -.03 .03 .4 .4 .5 .5 1 ];
UpperSidelobeGoal_dB=[-55 -30 -30 -39.5 -60 -60 -42.1 -55];
LowerSidelobeGoal_u= [ -1 1];
LowerSidelobeGoal_dB=[-80 -80];
AutomaticallyExemptMainBeam=1; % 1 to exempt main beam
from goals
NPatternPoints=512;
%%%%%%%%%%%%%%%%%%%%%%%%%%%%%%%%%%%%%%%%
% Optimization Parameters
%%%%%%%%%%%%%%%%%%%%%%%%%%%%%%%%%%%%%%%%
Niterations=1000;
Npopulation=50;
Phi1=2;
Phi2=2;
W=.4;
VRMSmax=.3;
Norder=2*Nelements;
```

```
%%%%%%%%%%%%%%%%%%%%%%%%%%%%%%%%%%%%%
% Miscellaneous Initialization
%%%%%%%%%%%%%%%%%%%%%%%%%%%%%%%%%%%%%
MagRange=MagMax-MagMin;
PhsRange=PhsMax-PhsMin;
SineTheta=[-1:2/NPatternPoints:1-2/NPatternPoints];
CosineTheta=sqrt(1-SineTheta.^2);
LowerEnvelope_dB=interp1(LowerSidelobeGoal_
u+[0:size(LowerSidelobeGoal_u,2)-1]*eps, LowerSidelobeGoal_
dB, SineTheta);
UpperEnvelope_dB=interp1(UpperSidelobeGoal_
u+[0:size(UpperSidelobeGoal_u,2)-1]*eps, UpperSidelobeGoal_
dB, SineTheta);
LowerEnvelope=10.^(LowerEnvelope_dB/20)';
UpperEnvelope=10.^(UpperEnvelope_dB/20)';
ElementWgtPower=CosineTheta.^CosineElementFactorExponent;
ElementWgtVoltage=sqrt(ElementWgtPower)';
tic;


%%%%%%%%%%%%%%%%%%%%%%%%%%%%%%%%%%%%%
% Initialize Particle Swarm
%%%%%%%%%%%%%%%%%%%%%%%%%%%%%%%%%%%%%
Swarm=rand(Norder,Npopulation);
SwarmVelocity=rand(Norder,Npopulation)*2-1;
%%%%%%%%%%%%%%%%%%%%%%%%%%%%%%%%%%%%%
% Score Particles
%%%%%%%%%%%%%%%%%%%%%%%%%%%%%%%%%%%%%
for kk=1:Npopulation
  MagWgts=Swarm(1:Nelements,kk).*MagRange+MagMin;
  PhsWgts=Swarm(Nelements+1:2*Nelements,kk).*PhsRange+Phs
  Min;
  ComplexWgts=(MagWgts.*exp(i*PhsWgts));
  SwarmScores(kk)=simpleCostFunction(...
    ComplexWgts,ElementWgtVoltage,LowerEnvelope,UpperEnvel
    ope,...
    AutomaticallyExemptMainBeam,EnforceSymmetry);
end
%%%%%%%%%%%%%%%%%%%%%%%%%%%%%%%%%%%%%
% Get Ready to Optimize
%%%%%%%%%%%%%%%%%%%%%%%%%%%%%%%%%%%%%
SwarmScoreMemory=SwarmScores;
SwarmLocalBest=Swarm;
GlobalBestScore=min(SwarmScores);
IndexOfBest=find(SwarmScores==GlobalBestScore);
SwarmGlobalBest=Swarm(:,IndexOfBest(1));


%%%%%%%%%%%%%%%%%%%%%%%%%%%%%%%%%%%%%
% Main Optimization Loop
%%%%%%%%%%%%%%%%%%%%%%%%%%%%%%%%%%%%%
```

```
for ii=1:Niterations
  %%%%%%%%%%%%%%%%%%%%%%%%%%%%%%%%%%%%
  % For Each Particle:
  %%%%%%%%%%%%%%%%%%%%%%%%%%%%%%%%%%%%
  for kk=1:Npopulation
    %%%%%%%%%%%%%%%%%%%%%%%%%%%%%%%%%%%%%
    % Update Velocity
    %%%%%%%%%%%%%%%%%%%%%%%%%%%%%%%%%%%%%
    SwarmSelfKnowledge=Phi1*rand(Norder,1).*...
      (SwarmLocalBest(:,kk)-Swarm(:,kk));
    SwarmSocialKnowledge=Phi2*rand(Norder,1).*...
      (SwarmGlobalBest-Swarm(:,kk));
    SwarmVelocity(:,kk)=W*SwarmVelocity(:,kk)+...
      SwarmSelfKnowledge+SwarmSocialKnowledge;
    VRMS=sqrt(sum(SwarmVelocity(:,kk).^2));
    if VRMS>VRMSmax
      SwarmVelocity(:,kk)=SwarmVelocity(:,kk)/VRMS*VRMSmax;
    end
    %%%%%%%%%%%%%%%%%%%%%%%%%%%%%%%%%%%%%
    % Update Position
    %%%%%%%%%%%%%%%%%%%%%%%%%%%%%%%%%%%%%
    Swarm(:,kk)=Swarm(:,kk)+SwarmVelocity(:,kk);
    Swarm(:,kk)=max(min(Swarm(:,kk),1),0);
    %%%%%%%%%%%%%%%%%%%%%%%%%%%%%%%%%%%%%
    % Update Score
    %%%%%%%%%%%%%%%%%%%%%%%%%%%%%%%%%%%%%
    MagWgts=Swarm(1:Nelements,kk).*MagRange+MagMin;
    PhsWgts=Swarm(Nelements+1:2*Nelements,kk).*PhsRange+Phs
    Min;
    ComplexWgts=(MagWgts.*exp(i*PhsWgts));
    SwarmScores(kk)=simpleCostFunction(...
      ComplexWgts,ElementWgtVoltage,LowerEnvelope,UpperEnvel
      ope,...
      AutomaticallyExemptMainBeam,EnforceSymmetry);
    %%%%%%%%%%%%%%%%%%%%%%%%%%%%%%%%%%%%%
    % Update Particle Memory
    %%%%%%%%%%%%%%%%%%%%%%%%%%%%%%%%%%%%%
    if SwarmScores(kk)<SwarmScoreMemory(kk)
      SwarmScoreMemory(kk)=SwarmScores(kk);
      SwarmLocalBest(:,kk)=Swarm(:,kk);
    end
    %%%%%%%%%%%%%%%%%%%%%%%%%%%%%%%%%%%%%
    % Update Group Knowledge
    %%%%%%%%%%%%%%%%%%%%%%%%%%%%%%%%%%%%%
    if SwarmScores(kk)<GlobalBestScore
      SwarmGlobalBest=Swarm(:,kk);
      GlobalBestScore=SwarmScores(kk);
    end
```

```
  end
  %%%%%%%%%%%%%%%%%%%%%%%%%%%%%%%%%%%%%%%
  % Save Performance Results
  %%%%%%%%%%%%%%%%%%%%%%%%%%%%%%%%%%%%%%%
  SaveSwarmAverageScore(ii)=mean(SwarmScores);
  SaveSwarmBestScores(ii)=GlobalBestScore;
end

%%%%%%%%%%%%%%%%%%%%%%%%%%%%%%%%%%%%%%%%%
% Calculate Best Pattern
%%%%%%%%%%%%%%%%%%%%%%%%%%%%%%%%%%%%%%%%%
NPatternPointsFine=NPatternPoints*8;
SineThetaFine=[-1:2/NPatternPointsFine:1-2/
NPatternPointsFine];
CosineThetaFine=sqrt(1-SineThetaFine.^2);
ElementWgtPowerFine=CosineThetaFine.^CosineElementFactorExpo
nent;
ElementWgtVoltageFine=sqrt(ElementWgtPowerFine).';
ComplexWgts=(SwarmGlobalBest(1:Nelements).*MagRange+Mag
Min).*...
  exp(i*(SwarmGlobalBest(Nelements+1:2*Nelements).*PhsRange+
  PhsMin));
if EnforceSymmetry
  ComplexWgts(Nelements/2+1:Nelements)=flipud(ComplexWgts(1:
  Nelements/2));
end
ComplexWgts=ComplexWgts/max(abs(ComplexWgts));
BestPattern=fftshift(ff t(ComplexWgts,NPatternPointsFine)).
*ElementWgtVoltageFine;
BestPattern_dB=20*log10(abs(BestPattern)+eps);
BestPatternNorm_dB=BestPattern_dB-max(BestPattern_dB);
ElapsedTimeMinutes=toc/60;

%%%%%%%%%%%%%%%%%%%%%%%%%%%%%%%%%%%%%%%%%
% Plot Results
%%%%%%%%%%%%%%%%%%%%%%%%%%%%%%%%%%%%%%%%%
figure
figscale=70; figoffsetx=20; figoffsety=20;
set(gcf,'Position',[figoffsetx figoffsety round(11.25*figsca
le+figoffsetx) round(6.75*figscale+figoffsety)])
fontsize=12;
subplot(2,2,1)
set(gca,'FontSize',fontsize)
plot([1:Niterations],SaveSwarmBestScores,[1:Niterations],Sav
eSwarmAverageScore,'LineWidth',2)
xlabel('Iteration #')
ylabel('Cost Function (arbitrary units)')
legend('Best','Average')
```

```
axis tight
subplot(2,2,2)
[ax,h1,h2]=plotyy([1:Nelements],abs(ComplexWgts),[1:Nelement
s],angle(ComplexWgts)-min(angle(ComplexWgts))));
set(h1,'Marker','.','LineStyle','none');
set(h2,'Marker','.','LineStyle','none');
axes(ax(1));
set(gca,'FontSize',fontsize)
ylabel('Amplitude (Volts) (blue)')
ylim([0 1])
axes(ax(2));
set(gca,'FontSize',fontsize)
ylabel('Phase (radians) (green)')
title('Aperture weights from Particle Swarm')
subplot(2,1,2)
set(gca,'FontSize',fontsize)
plot(SineThetaFine,BestPatternNorm_dB,'LineWidth',2);
hold on
plot(LowerSidelobeGoal_u, LowerSidelobeGoal_dB,
'r-.','LineWidth',2)
plot(UpperSidelobeGoal_u, UpperSidelobeGoal_dB,
'r:','LineWidth',2)
axis tight
ylim([-80 0])
grid on
title(['Far field Pattern (Target sidelobe constraints in
red). Elapsed time = ' num2str(ElapsedTimeMinutes) '
minutes.'])
ylabel('Magnitude (dB)')
xlabel('Sin(\theta)')
```

References

Boeringer, D. W., and D. H. Werner. Particle Swarm Optimization versus Genetic Algorithms for Phased Array Synthesis. *IEEE Transactions on Antennas and Propagation*, 2004: 771–779.

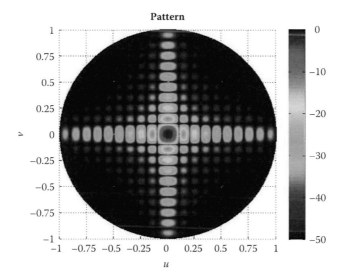

Color Figure 2.19 Boresite antenna pattern (no electronic scan) in radar coordinates.

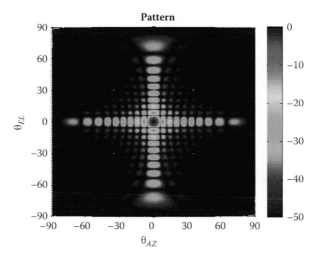

Color Figure 2.20 Boresite antenna pattern (no electronic scan) in sine space.

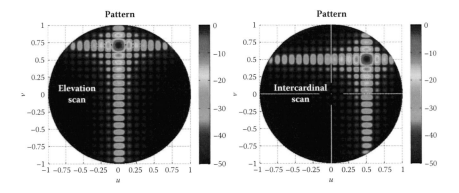

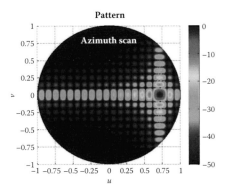

Color Figure 2.21 Electronically scanned antenna patterns in the principal planes ($\theta_o = 60°$, $\phi_o = 0°$ and $\theta_o = 60°$, $\phi_o = 90°$) and intercardinal plane ($\theta_o = 60°$, $\phi_o = 45°$).

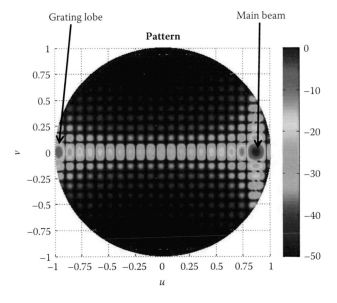

Color Figure 2.22 Electronic scan beyond 60° allows the fully formed grating lobe to appear in real space.

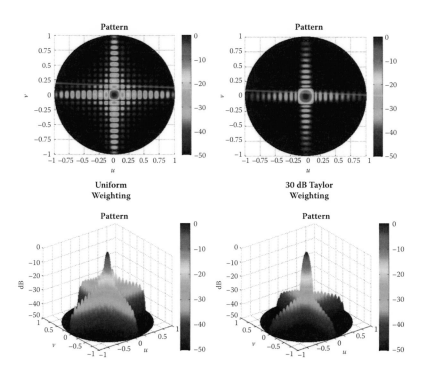

Color Figure 2.23 SLL reduction using a 30 dB Taylor weighting compared to a uniform distribution.

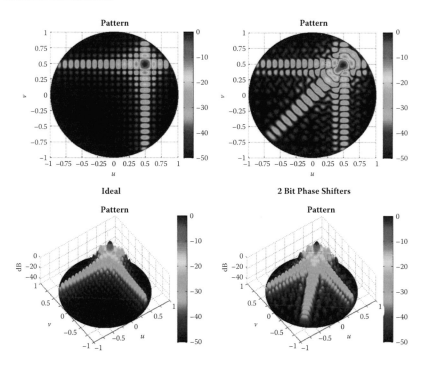

Color Figure 2.25 Pattern from Figure 2.20. Boresite antenna pattern (no electronic scan) in sine space with 2-bit phase shifters and attenuators and a pattern with 6-bit phase shifters and attenuators.

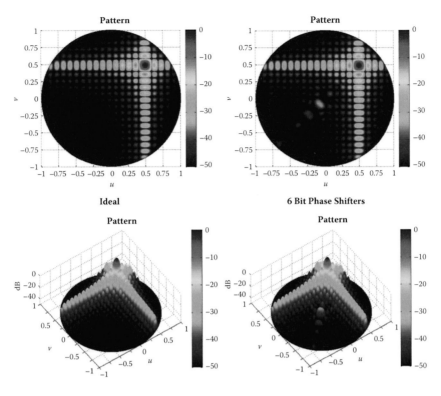

Color Figure 2.26 Comparison of an ideal pattern and a pattern with 6-bit phase shifters and attenuators.

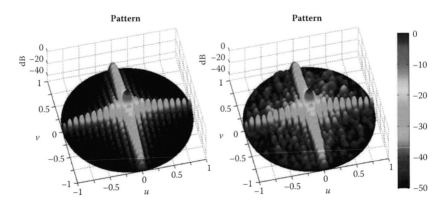

Color Figure 2.28 Comparison of an ideal pattern with a pattern that has the random amplitude and phase errors shown in Figure 2.27.

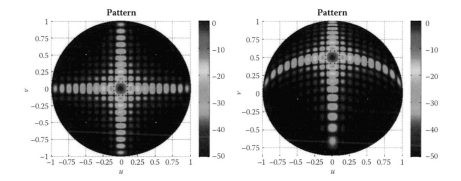

Color Figure 2.32 Effects of pitch on an ESA pattern.

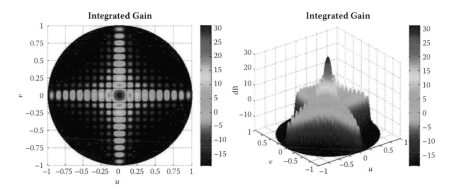

Color Figure 2.33 Integrated gain of the pattern in Figure 2.20 Boresite antenna pattern (no electronic scan) in sine space.

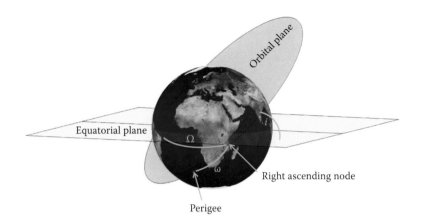

Color Figure 5.5 Orbital orientation.

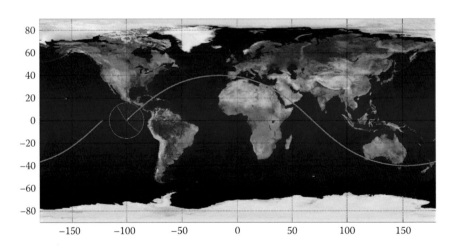

Color Figure 5.11 Horizon and FOV projected to map.

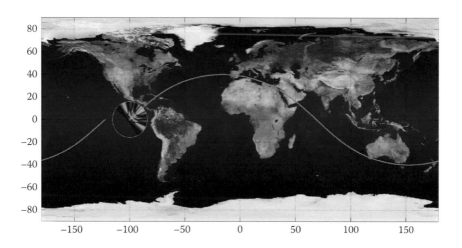

Color Figure 5.13 Antenna pattern projected to map.

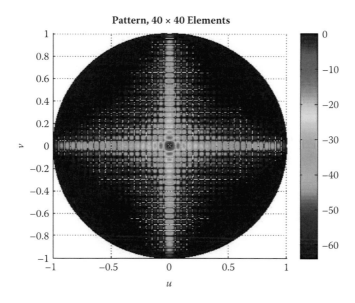

Color Figure 6.5 Antenna pattern for 40×40 grid of elements spaced $l/2$ apart. The pattern has been normalized to the peak of the beam.

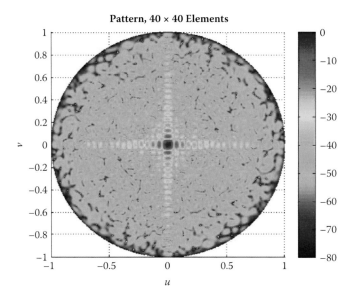

Color Figure 6.9 Two-dimensional pattern of antenna with 10% failures.

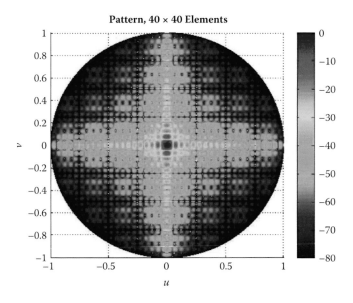

Color Figure 6.12 Two-dimensional pattern from the array with a failed subarray as described in Figure 11.

chapter five

Spaceborne Application of Electronically Scanned Arrays

Timothy Cooke
Northrop Grumman Electronic Systems

Contents

5.1   Introduction

Electronically scanned arrays (ESAs) have been used in spaceborne applications since 1978 when NASA began the SEASAT mission. SEASAT was the first synthetic aperture radar (SAR) mission in space, and used a horizontally polarized, L-band ESA to map the earth's surface. After SEASAT, NASA conducted three additional SAR missions in the 1980s called the shuttle imaging radar (SIR) missions (SIR-A, SIR-B, and SIR-C). In these missions, ESA-based SAR payloads were operated from the cargo bay of the space shuttle. Currently there are several spaceborne SAR payloads that use ESAs, including Germany's TerraSAR-X, Canada's RADARSAT-1 and RADARSAT-2, the European Space Agency's Envisat, and the European remote sensing satellite (ERS).

In this chapter, we present a set of tools for analyzing and visualizing spaceborne ESA applications. This section will illustrate a method for determining the ESA field of view (FOV) as the intersection of the ESA scan volume (the angles to which the ESA can electronically scan without grating lobes) with the ESA perspective of the earth's horizon. A method for projecting the field of view, as well as two-dimensional antenna patterns to the earth's surface, is developed. A simple geometry is constructed by modeling an ideal two-body orbit about a spherical earth.

5.2   Two-Body Orbit Propagation

In order to evaluate ESAs in the context of space, it is helpful to model an orbital geometry. A brief overview of orbital mechanics is followed with a simplified presentation of the two-body orbit and Kepler's equations. For the purposes of visualizing scan volumes and antenna patterns, several simplifications can be made to the geometry with little impact to the results. In this discussion we will consider a spherical earth, and we will neglect orbital perturbations.

Kepler's first law states that all orbits are conic sections (ellipses, circles, hyperbolas, or parabolas). Elliptical and circular orbits periodically repeat, while hyperbolic and parabolic orbits do not repeat. We will focus on elliptical orbits, with circular orbits being a special case. All conic sections can be defined by three parameters, a, c, and p. Figure 5.1 shows an ellipse with these parameters labeled.

The semimajor axis, a, and the semiminor axis, b, of an ellipse define the eccentricity, e, as shown in Equation (5.1). The eccentricity of an ellipse

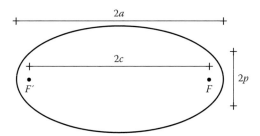

Figure 5.1 Conic section parameters. See insert.

is between zero and unity. A circle is an ellipse with equal semimajor and semiminor axes, and ellipticity of 0.

$$e = \frac{\sqrt{a^2 - b^2}}{a} \qquad (5.1)$$

The semiparameter, p, is the distance of a line perpendicular to the semimajor axis connecting the focus to the ellipse, and is defined in Equation (5.2).

$$p = a(1 - e^2) \qquad (5.2)$$

In the case of an earth-orbiting satellite, the earth is considered the central body and lies at the center of a circular orbit, or the primary focus of an elliptical orbit. The points at which the orbiting satellite is nearest and farthest from the earth's center of mass are the perigee and apogee, shown respectively in Equations (5.3) and (5.4):

$$r_p = alt_{min} + R_{earth} \qquad (5.3)$$

$$r_a = alt_{max} + R_{earth} \qquad (5.4)$$

These distances are equal to the sum of the earth's radius and the satellite altitude. The mean spherical radius of the earth is approximately 6,371 km. The semimajor and semiminor axes are determined by the distance to perigee and apogee, as shown in Equations (5.5) and (5.6), providing a convenient and simple means of specifying an elliptical orbit.

$$a = \frac{r_p + r_a}{2} \qquad (5.5)$$

$$b = \sqrt{r_p \cdot r_a} \qquad (5.6)$$

The orbital period is the time required for the satellite to complete one full revolution around the earth. The orbital period is defined by the semimajor axis, a, and the gravitational constant, μ, as shown in Equations (5.7) and (5.8).

$$T = 2\pi \sqrt{\frac{a^3}{\mu}} \qquad (5.7)$$

$$\mu = G(m_{earth} + m_{sat}) = 3.986004415 \times 10^5 \, \frac{km^3}{s^2} \qquad (5.8)$$

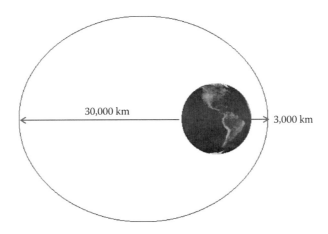

Figure 5.2 Elliptical orbit example.

The example in Figure 5.2 shows an orbital ellipse with perigee and apogee distances of 3,000 and 30,000 km, respectively. Using the equations mentioned previously, the semimajor and semiminor axes are found to be 22,871 and 18,462 km, respectively, with an eccentricity of 0.59. This figure is generated by the example script main_example1.m in Section 5.6.

The position of the satellite in a two-body orbit is defined by the distance, r, from the primary focus, and by the angle from the line connecting the focus and the perigee, v, known as the true anomaly. These parameters are illustrated in Figure 5.3. Kepler's second law states that a line

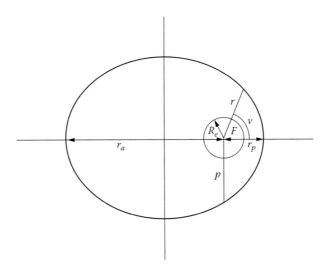

Figure 5.3 Orbital position parameters.

drawn from the central body to the orbiting body sweeps out equal areas in equal times. This important result was the foundation of Kepler's equation, which relates time and position within an orbit. Kepler's equation is stated as

$$M = n(t - T) = E - e\sin(E) \tag{5.9}$$

where M is the mean anomaly, n is the mean motion, E is the eccentric anomaly, and $(t - T)$ is the time of flight from perigee. Mean motion, n, is the mean angular velocity through the orbit

$$n = \sqrt{\frac{\mu}{a^3}} \tag{5.10}$$

Mean anomaly is the position based on this mean motion and the time of flight.

Consider a circle with radius equal to the semimajor axis of the ellipse and concentric with the ellipse. While the true anomaly describes the angular position of the satellite along the ellipse, the eccentric anomaly, E, is the angular position of a point on the circle with equal horizontal displacement from the focus. Figure 5.4 illustrates the relationship between the true anomaly and the eccentric anomaly.

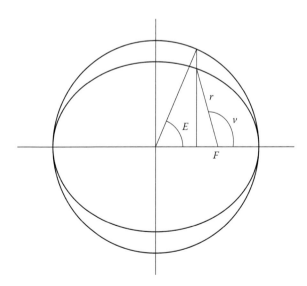

Figure 5.4 True anomaly and eccentric anomaly.

Vallado (2007) presents a numerically efficient algorithm to solve Kepler's equation using Newton's method. We will use this algorithm to solve for E. The following equations relate v and r to E:

$$2\tan(v) = \sqrt{\frac{1+e}{1-e}}\tan\left(\frac{E}{2}\right) \tag{5.11}$$

$$r = a(1 - e\cos(E)) \tag{5.12}$$

The polar coordinates v and r are converted to Cartesian coordinates as follows:

$$x = r\cos(v)$$
$$y = r\sin(v) \tag{5.13}$$

These coordinates describe the position in the orbital plane at a given time. In addition to position, we can also solve for velocity using the following equations:

$$v_x = -\sqrt{\frac{\mu}{p}}\sin v$$

$$v_y = \sqrt{\frac{\mu}{p}}(e + \cos v) \tag{5.14}$$

The solution of Kepler's equation for position and velocity of a satellite about a central body as a function of time of flight is a useful tool for generating a realistic orbital geometry. This geometry can be used for time-varying analysis of space-based ESA applications. The material in this section has been encapsulated in the defineOrbit.m MATLAB® function in Section 5.6. The defineEarth.m function defines several earth-specific constants and loads a base map that is used to generate the graphics in the following sections.

5.3   Coordinate Systems

Several coordinate systems are used in this chapter. In this section, we focus on the pertinent coordinate systems, and the transformations from one to another. The two-body orbit described in the previous section was defined in a plane. In order to simulate a realistic orbit, the orbital plane must be rotated into an orientation about the earth. In astrodynamics, this orientation is typically described by three Euler angles. These angles are

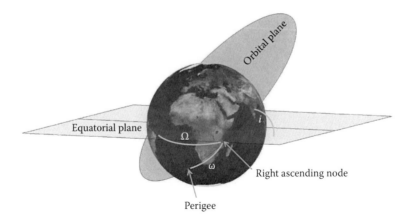

Figure 5.5 Orbital orientation. See insert.

the right ascension of the ascending node Ω, inclination i, and the argument of perigee ω. Ω is the longitude that the satellite crosses the earth's equatorial plane from south to north. Inclination is the angle between the orbital plane and the earth's equatorial plane. The argument of perigee is the angular distance between the node and perigee within the orbital plane. These angles are shown in Figure 5.5.

The two-body orbit described in the previous section assumed an inertial (nonaccelerating) reference frame. When modeling satellite orbits around earth, it is convenient to define an inertial coordinate system with earth's center at the origin. This coordinate system is called the earth-centered inertial (ECI) frame. Most ECI coordinate systems align the x axis with the vernal equinox, and the z axis with the north pole. The x-y axis is the equatorial plane. The rotation of the orbital plane by Ω, i, and ω transforms the orbit to ECI coordinates.

Earth-centered, earth-fixed (ECEF) coordinates also define the origin at the center of the earth, but the axes are fixed to the earth's surface. The x axis is defined from the origin through the intersection of the equator and prime meridian. The z axis extends from the origin through the north pole (true north). The y axis also points out the equator, 90° from the x axis. ECEF coordinates are useful for specifying ground locations, since the coordinates are fixed to the earth. Terrestrial locations are also commonly specified using latitude and longitude. For a spherical earth, latitude and longitude are computed from ECEF using a Cartesian to spherical transformation (cart2sph in MATLAB).

Every space vehicle uses several coordinate systems to describe the attitude (orientation) of the vehicle, and each of its various sensors and systems. For the purpose of this discussion, we will define two coordinate systems to describe the relative attitude of the space vehicle and the

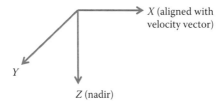

Figure 5.6 Satellite coordinate system.

ESA: space vehicle coordinates (denoted SAT) and antenna coordinates (denoted ANT). It is assumed that both systems have the same origin at the phase center of the ESA. The space vehicle coordinates will be defined with x axis aligned with the space vehicle's velocity vector, y axis to the right of the vehicle's direction of travel, and z axis down from the space vehicle to the earth's center (nadir vector). See Figure 5.6. The antenna coordinate system is used for generating antenna patterns and computing angles (azimuth and elevation) relative to the antenna. The antenna coordinate system defined in Figure 5.7 assumes the array is in the x–y plane, with its phase center at the origin, facing in the positive z direction (also known as boresite). Azimuth and elevation angles are defined in Figure 5.7. The transformation from space vehicle coordinates to antenna coordinates determines the mechanical orientation of the ESA on the space vehicle.

Sine space, as mentioned earlier in Chapter 2, is a natural coordinate system for antenna systems. It collapses the three-dimensional space in

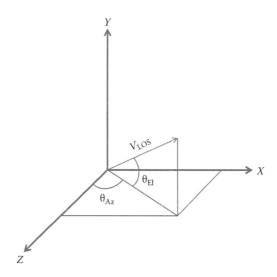

Figure 5.7 Antenna coordinate system.

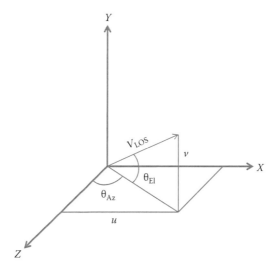

Figure 5.8 Sine space coordinate system.

front of the antenna into two dimensions. We will use sine space to visualize the earth's horizon as seen by the ESA, and to compute its intersection with the ESA scan volume to determine the resulting field of view. Figure 5.8 shows the u and v components of a unit line of sight vector. A unit vector, v_{ECEF}, from the ESA to a point on the earth can be converted from ECEF coordinates to sine space as follows:

$$v_{ANT} = T_{SAT2ANT} \cdot (T_{ECEF2SAT} v_{ECEF}) = \begin{bmatrix} v_u \\ v_v \\ v_w \end{bmatrix}$$

(5.15)

$$v_{SS} = \begin{bmatrix} v_u \\ v_v \end{bmatrix}$$

It is often useful to find the point of intersection of a line of sight vector from the ESA to the earth. By modeling the earth as a sphere, we can find this intersection by solving a quadratic equation. There can be two solutions (the vector intersects the sphere at two points), one solution (the vector is tangent to the sphere), or no solutions (the vector does not intersect the sphere). If v is the line of sight vector and c is the position of the ESA (both in ECEF coordinates), the quadratic equation in Equation (5.16) solves for the distance to the intersection points. Since we're interested in finding points on the earth visible to the ESA, we are always interested in the minimum distance, since by definition the larger distance is on the opposite side of the planet. If d is undefined or imaginary, then the vector does not intersect the earth. The position of the intersection point on the earth in ECEF coordinates is found by multiplying the unit vector v by the distance and adding the position of the ESA. The los2ecef.m function in Section 5.6 uses this method to compute the intersection of a line of sight vector with the earth.

$$d = -v^T c \pm \sqrt{(v^T c)^2 - c^T c + R_{Earth}^2}$$

$$d = \min[d_1, d_2] \tag{5.16}$$

$$p_{ECEF} = dv + c$$

5.4   Computing Field of View

As discussed previously, ESAs will exhibit grating lobes when scanned beyond the limits imposed by the element spacing. The region over which the ESA can be scanned without grating lobes is called the scan volume. Even if the scan volume is not limited by element spacing, it is often limited by the element pattern. If the ESA has uniform spacing in both x and y dimensions, the resulting FOV can be approximated as a circle of radius $\sin(\theta_{max})$ in sine space. If the ESA has different spacing in the x and y dimensions, the scan volume can be approximated as an ellipse with radii $\sin(\theta_{az,max})$ and $\sin(\theta_{el,max})$, respectively. Figure 5.9 shows an example of an elliptical scan volume in sine space.

For earth-observing satellites, the earth's horizon relative to the ESA presents an additional pointing constraint. The distance to the visible horizon is computed by first finding the earth angle to the horizon.

$$\theta_{hor} = \cos^{-1}\left(\frac{R_{Earth}}{alt + R_{Earth}}\right)$$

$$R_{hor} = (alt + R_{Earth})\sin\theta_{hor} \tag{5.17}$$

The horizon on a spherical earth is a circle with a constant distance of R_{hor} from the satellite. This circle is defined in geometry as a small circle. The computeHorizon.m function in Section 5.6 computes the coordinates (sine space, latitude/longitude, and ECEF) of the horizon for a given space vehicle position and orientation.

The combination scan volume constraints and the horizon result in the field of view (FOV). The field of view is the area of the earth's surface

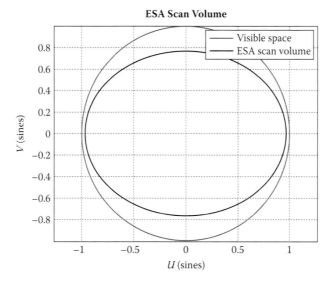

Figure 5.9 Scan volume example in sine space.

that can be covered by the ESA. The field of view is the intersection of the scan volume and the horizon in sine space as shown in Figure 5.10. The computeFOV.m function in Section 5.6 computes this intersection and converts the FOV to sine space and latitude/longitude. Figure 5.11 shows the horizon and FOV of Figure 5.10 projected to ECEF coordinates.

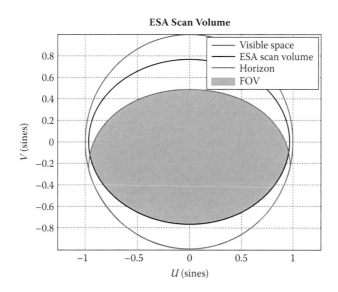

Figure 5.10 Projection of horizon to sine space.

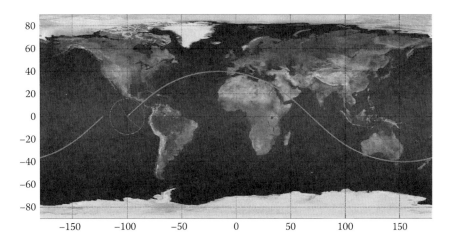

Figure 5.11 Horizon and FOV projected to map. See insert.

5.5   Projecting Antenna Patterns to Geodetic Coordinates

In addition to visualizing the FOV, it is often useful to visualize the projection of the ESA's antenna pattern to the ground. One method to project antenna patterns to the earth's surface is to define a grid of points within the field of view, and to transform these points from longitude and latitude or ECEF to sine space. A sine space antenna pattern (Figure 5.12) interpolated to these points can then be directly displayed on a map. Figure 5.13 shows

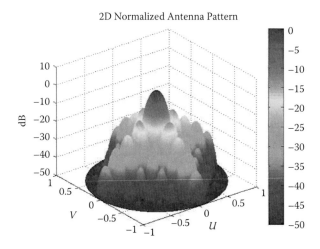

Figure 5.12 Sine space antenna pattern.

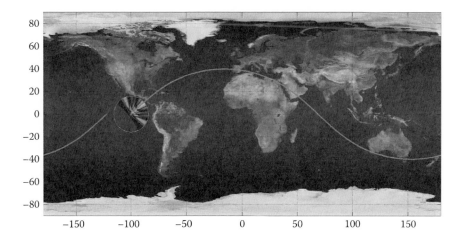

Figure 5.13 Antenna pattern projected to map. See insert.

an example antenna pattern projected to the earth's surface. The code to generate these plots is included in Section 5.6 under main_example3.m.

5.6 MATLAB Program Listings

This section contains a listing of all MATLAB programs and functions used in this chapter.

5.6.1 defineOrbit.m

```
function Sat = defineOrbit(Earth,altMin,altMax,inclination,O
MEGA,omega,nRevs,offsetRevs,timeStep)

% parse inputs
Sat.altPerigee = altMin; % min. altitude (km)
Sat.altApogee = altMax; % max altitude (km)
Sat.inclination = inclination; % inclination angle (deg)
Sat.OMEGA = OMEGA; % right ascension of the ascending node,
RAAN (deg)
Sat.omega = omega; % argument of perigee, measured from
RAAN (deg)

% calculate orbital parameters
Sat.rp = Earth.Re_km + Sat.altPerigee; % perigee distance (km)
Sat.ra = Earth.Re_km + Sat.altApogee; % apogee distance (km)
Sat.a = (Sat.rp+Sat.ra)/2; % semi-major axis (km)
Sat.b=sqrt(Sat.rp*Sat.ra); % semi-minor axis (km)
```

```
Sat.e = sqrt(Sat.a^2 - Sat.b^2)/Sat.a; % eccentricity
(unitless)
Sat.p = Sat.a*(1 - Sat.e^2); % semi-parameter (km)
Sat.n = sqrt(Earth.mu/Sat.a.^3); % mean motion (rad/sec)
Sat.T = 2*pi*sqrt(Sat.a^3/Earth.mu); % orbital period (sec)

% define time vector
Sat.nRevs = nRevs; % number of revolutions
Sat.nRevOffset = offsetRevs; % offset from t=0 in revolutions
Sat.time = Sat.nRevOffset*Sat.T + 0:timeStep:Sat.T*Sat.
nRevs; % time (sec)
% loop over time
for ii = 1:length(Sat.time)
   t = Sat.time(ii); % current time (sec)

   % solve for current position in orbit (Newton's Method,
   Vallado)
   tol = 10^-8; % set tolarance
   M = Sat.n*t; % compute mean anomaly, M (rad)
   M = mod(M,2*pi); % restrict M to [0,2*pi]
   if M > pi
      Etemp = M - Sat.e; % initialize eccentric anomaly, E
   else
      Etemp = M + Sat.e;
   end
   delta = 1;
   while abs(delta) > tol % iterate to solve for eccentric
   anomaly, E
      fE = M - Etemp + Sat.e * sin(Etemp);
      fEprime = -1 + Sat.e * cos(Etemp);
      delta = fE/fEprime;
      if abs(delta) > tol % iterate until delta < tolerance
         Etemp = Etemp - delta; % update estimate of E
      else
         E = Etemp;
      end
   end

   % solve for true anomaly, nu (rad) and radial distance r
   (km)
   nu = 2*atan2(sqrt(1+Sat.e)*sin(E/2),sqrt(1-
   Sat.e)*cos(E/2)); % (rad)
   r_km = Sat.a*(1-Sat.e*cos(E)); % (km)

   % convert position from polar to cartesian coordinates
   x_km = r_km.*cos(nu);
   y_km = r_km.*sin(nu);
   z_km = 0;
```

```
% solve for current velocity
vx_kmpersec = -sqrt(Earth.mu/Sat.p)*sin(nu); % (km/sec)
vy_kmpersec = sqrt(Earth.mu/Sat.p)*(Sat.e+cos(nu)); %
(km/sec)
vz_kmpersec = 0; % (orbit defined in the x-y plane) (km/sec)

% rotate orbital plane by omega, inclination, and w (ECI
coordinates)
W = Sat.OMEGA*pi/180; % (rad)
inc = Sat.inclination*pi/180; % (rad)
w = Sat.omega*pi/180; % (rad)
Mrot = [cos(W) * cos(w) - sin(W) * cos(inc) * sin(w)
-cos(W) * sin(w) - sin(W) * cos(inc) * cos(w) sin(W) *
sin(inc);
    sin(W) * cos(w) + cos(W) * cos(inc) * sin(w) -sin(W) *
    sin(w) + cos(W) * cos(inc) * cos(w) -cos(W) *
    sin(inc);
    sin(inc) * sin(w) sin(inc) * cos(w) cos(inc);];
posECI = Mrot*[x_km; y_km; z_km]; % rotate position
vector into ECI coordinates
Sat.x_km_ECI(ii) = posECI(1);
Sat.y_km_ECI(ii) = posECI(2);
Sat.z_km_ECI(ii) = posECI(3);
velECI = Mrot*[vx_kmpersec;vy_kmpersec;vz_kmpersec]; %
rotate velocity vector into ECI coordinates
Sat.vx_kmpersec_ECI(ii) = velECI(1);
Sat.vy_kmpersec_ECI(ii) = velECI(2);
Sat.vz_kmpersec_ECI(ii) = velECI(3);
Sat.v_kmpersec(ii) = sqrt(Earth.mu*(2./r_km(end)-1/
Sat.a)); % solve for velocity magnitude

% rotate ECI orbital position by Earth's rotation to get
ECEF position
Sat.x_km(ii) = Sat.x_km_ECI(ii)*cos((t+Earth.
rotTimeOffset)*Earth.rotRate*pi/180) + Sat.y_km_
ECI(ii)*sin((t+Earth.rotTimeOffset)*Earth.
rotRate*pi/180);
Sat.y_km(ii) = Sat.x_km_ECI(ii)*-sin((t+Earth.
rotTimeOffset)*Earth.rotRate*pi/180) + Sat.y_km_
ECI(ii)*cos((t+Earth.rotTimeOffset)*Earth.
rotRate*pi/180);
Sat.z_km(ii) = Sat.z_km_ECI(ii);

% rotate ECI orbital velocity by Earth's rotation to get
ECEF velocity
Sat.vx_kmpersec(ii) = Sat.vx_kmpersec_
ECI(ii)*cos((t+Earth.rotTimeOffset)*Earth.
rotRate*pi/180)+Sat.vy_kmpersec_ECI(ii)*sin((t+Earth.
rotTimeOffset)*Earth.rotRate*pi/180);
```

```matlab
Sat.vy_kmpersec(ii) = Sat.vx_kmpersec_ECI(ii)*-
sin((t+Earth.rotTimeOffset)*Earth.rotRate*pi/180)+Sat.
vy_kmpersec_ECI(ii)*cos((t+Earth.rotTimeOffset)*Earth.
rotRate*pi/180);
Sat.vz_kmpersec(ii) = Sat.vz_kmpersec_ECI(ii);

% convert ECEF to latitude/longitude/altitude (assuming
spherical earth)
[TH,PHI] = cart2sph(Sat.x_km(ii),Sat.y_km(ii),Sat.z_
km(ii));
Sat.lat(ii) = PHI*180/pi; % latitude (deg)
Sat.lon(ii) = TH*180/pi; % longitude (deg)
Sat.Altitudekm(ii) = sqrt(Sat.x_km(ii).^2+Sat.y_
km(ii).^2+Sat.z_km(ii).^2)-Earth.Re_km; % altitude (km)

% define current nadir vector (ECI)
nadirvector = [-Sat.x_km_ECI(ii) -Sat.y_km_ECI(ii)
-Sat.z_km_ECI(ii)];
nadirvector = nadirvector/norm(nadirvector);

% define current velocity vector (ECI)
velocityvector = [Sat.vx_kmpersec_ECI(ii) Sat.vy_
kmpersec_ECI(ii) Sat.vz_kmpersec_ECI(ii)];
velocityvector = velocityvector/norm(velocityvector);

% define satellite local level coordinate system
ux = velocityvector;
uz = nadirvector;
uy = cross(uz,ux);
uy = uy/norm(uy);
ux = cross(uy,uz);

% define coordinate transforms from ECI (xyz) to local
level (sat)
Sat.xyz2sat(:,:,ii) = [ux(:),uy(:),uz(:)].';
Sat.sat2xyz(:,:,ii) = [ux(:),uy(:),uz(:)];
end

% save position, pos (ecef, km)
Sat.pos = [Sat.x_km; Sat.y_km; Sat.z_km];

% save position, llh (lat,lon,hgt)
Sat.llh = [Sat.lat; Sat.lon; Sat.Altitudekm + Earth.Re_km];

PLOT_FIGURE = 0;
if PLOT_FIGURE == 1
   % plot ground track
   figure;
   image(Earth.imgLon,Earth.imgLat,Earth.img);
```

```
   axis image; axis xy; grid on;
   hold on;
   plot(Sat.lon,Sat.lat,'g.');
end
return
```

5.6.2 defineEarth.m

```
function Earth = defineEarth
% - creates a structure of planetary constants for a
spherical Earth
% - also loads a copyright-free image of Earth from
Wikipedia

% planetary constants
Earth.Re_km = 6371; % mean spherical radius (km)
Earth.mu = 398600.4418; % (km^3/s^2)
Earth.Tsec = 86164.098903691; % 'stellar day' or period of
rotation relative to fixed stars (sec)
Earth.rot Rate = 360/Earth.Tsec; % Earth rotation rate,
(deg/sec)
Earth.rot Time Offset = 0; % Rotation time offset (sec)

% load Earth image from Wikipedia (copyright free image)
% [http://en.wikipedia.org/wiki/File:Earthmap1000x500compac.
jpg]
Earth.img = imread('Earthmap1000x500compac.jpg','jpeg'); %
whole earth image
Earth.imgLat = linspace(90,-90,size(Earth.img,1)); %
latitude vector (deg)
Earth.imgLon = linspace(-180,180,size(Earth.img,2)); %
longitude vector (deg)

% plot Earth image
PLOT_FIGURE = 0;
if PLOT_FIGURE == 1
   figure;
   image(Earth.imgLon,Earth.imgLat,Earth.img);
   axis image; axis xy; grid on;
   xlabel('Longitude'); ylabel('Latitude')
end
return
```

5.6.3 makeEllipse.m

```
function [x,y] = makeEllipse(r1,r2,n)
% returns n pairs of coordinates of an ellipse defined by
% semi-major/minor axes r1 and r2
```

```
a = max(r1,r2);
b = min(r1,r2);
theta = linspace(0,2*pi,n);
r = (a*b)./sqrt((b*cos(theta)).^2 + (a*sin(theta)).^2);
[x,y] = pol2cart(theta,r);
return
```

5.6.4   defineESA.m

```
function ESA = defineESA(azMech,elMech,azMaxScan,elMaxScan)
%% define mechanical orientation of array
% rotate = Azimuth rotation about z-axis (nadir)
DCMrotate = [cosd(azMech) -sind(azMech) 0;
   sind(azMech) cosd(azMech) 0;
   0 0 1];

% tilt = Elevation rotation about y-axis (cross-track)
DCMtilt = [cosd(elMech) 0 sind(elMech);
   0 -1 0;
   -sind(elMech) 0 cosd(elMech)];

% define coordinate transformation from satelite coordinates
to antenna
% coordinates as the combination of tilt/rotate DCMs
ESA.ant2sat = DCMtilt*DCMrotate;
ESA.sat2ant = ESA.ant2sat.';

%% define ESA scan volume as a circle or ellipse in sine space
[ESA.umax,ESA.vmax] = makeEllipse(sind(azMaxScan),sind(elMax
Scan),180);

% define unit circle (visible space)
[ESA.uunit,ESA.vunit] = makeEllipse(1,1,180);

PLOT_FIGURE = 1;
if PLOT_FIGURE == 1
   % plot scan volume and unit circle
   figure;
   plot(ESA.uunit,ESA.vunit,ESA.umax,ESA.vmax);
   axis equal; grid on;
   xlabel('U (sines)'); ylabel('V (sines)');
   title('ESA Scan Volume');
   legend('Visible Space','ESA Scan Volume')
end

%% generate a 2D pattern
Pattern 2D;
```

```
Esau=sinespace.umat;
Esau=sinespace.vmat;
Esa.pat=plotPAT;

PLOT_FIGURE = 1;
if PLOT_FIGURE == 1
   figure; surf(ESA.u,ESA.v,ESA.pat); colorbar;
   shading 'interp';
   title('2D Normalized Antenna Pattern');
   xlabel('U'); ylabel('V'); zlabel('dB');
end
return
```

5.6.5   los2ecef.m

```
function [pos,llh] = los2ecef(P1,Re,LOS)
% Finds the intersection of a ray along a LOS vector
launched from point P1
% toward a sphere with radius, Re

u = LOS/norm(LOS);
c = -P1; % center LOS at the origin, shift the center of
the sphere
d1 = u'*c + sqrt((u'*c)^2 - c'*c + Re^2); % quadratic
solution to find distance to intersection
d2 = u'*c - sqrt((u'*c)^2 - c'*c + Re^2); % distance to
second intersection point
d = min(d1,d2); % the shortest distance corresponds to the
intersection point
pos = d*u - c; % ecef position of intersection point

% if pos is not real, then the LOS doesn't intersect the
sphere
if ~isreal(pos)
   pos = nan(3,1);
   llh = nan(3,1);
else
   [lon,lat,hgt] = cart2sph(pos(1),pos(2),pos(3));
   llh = [lat*180/pi; lon*180/pi; hgt];
end
```

5.6.6   computeHorizon.m

```
function hor = computeHorizon(Earth,xyz2sat,sat2ant,pos,llh)

% compute earth angle to horizon
thetaEarth = 0.999 * acosd(Earth.Re_km/llh(3));
```

```
% compute range to horizon
rangeHor = llh(3) * sind(thetaEarth);

% define the small circle on the spherical earth that
describes the horizon
u1 = pos(:)/norm(pos);
u2 = cross(u1,[1;0;0]);
u2 = u2/norm(u2);
u3 = cross(u1,u2);
u3 = u3/norm(u3);
R = [u2,u3,u1].'; % rotation matrix that converts back to ecef

% define small circle in polar coordinates
npts = 200;
gr = sind(thetaEarth)*Earth.Re_km; % ground range to
horizon (km)
z1 = cosd(thetaEarth)*Earth.Re_km; % distance from earth
center to horizon-horizon chord (km)
theta = linspace(0,2*pi,npts);
[xhor,yhor,zhor] = pol2cart(theta,repmat(gr,1,npts),repmat(z
1,1,npts));
circ = [xhor;yhor;zhor]; % xyz coordinates of horizon circle
circ2 = R.'*circ; % rotate convenient xyz coordinates back
to ecef

% convert from ecef back to lat/lon
[lonHor,latHor] = cart2sph(circ2(1,:),circ2(2,:),circ2(3,:));
lonHor = lonHor * 180/pi;
latHor = latHor * 180/pi;

% convert to sine space
LOS = circ2 - repmat(pos,1,npts); % LOS vector to each
horizon point
range = sqrt(sum(LOS.^2,1)); % slant range to horizon
LOS = LOS./repmat(range,3,1); % convert LOS to unit vector
LOS = sat2ant * xyz2sat * LOS; % convert to antenna
coordinates
vHor = LOS(1,:);
uHor = LOS(2,:);
iNotVisible = sign(LOS(3,:)) == -1;
vHor(iNotVisible) = sign(vHor(iNotVisible)).*sqrt(1-
uHor(iNotVisible).^2);

hor.u = uHor;
hor.v = vHor;
hor.x = circ2(1,:);
hor.y = circ2(2,:);
hor.z = circ2(3,:);
hor.lon = lonHor;
```

```
hor.lat = latHor;
hor.range = rangeHor;
hor.thetaEarth = thetaEarth;
return
```

5.6.7 computeFOV.m

```
function fov = computeFOV(Earth,ESA,sat2xyz,ant2sat,pos,hor)
% Field of View (FOV) is the intersection of Scan Volume
and Horizon

% find the intersection of scan volume and horizon
in1 = inpolygon(hor.u,hor.v,ESA.umax,ESA.vmax);
in2 = inpolygon(ESA.umax,ESA.vmax,hor.u,hor.v);
uFOV = [hor.u(in1), ESA.umax(in2)];
vFOV = [hor.v(in1), ESA.vmax(in2)];
k = convhull(uFOV,vFOV); % use convex hull to reorder
points; assumes FOV should be convex
uFOV = uFOV(k);
vFOV = vFOV(k);

% convert FOV to lat/lon
latFOV = nan(size(uFOV));
lonFOV = nan(size(uFOV));
x = nan(size(uFOV));
y = nan(size(uFOV));
z = nan(size(uFOV));
zFOV = sqrt(1 - uFOV.^2 - vFOV.^2); % z component of LOS
vector in antenna coordinates
LOS = [vFOV;uFOV;zFOV];
LOS = sat2xyz * ant2sat * LOS; % convert LOS vector to
ecef coordinates
for ii = 1:size(LOS,2)
   [p,llh] = los2ecef(pos,Earth.Re_km,LOS(:,ii));
   latFOV(ii) = llh(1);
   lonFOV(ii) = llh(2);
   x(ii) = p(1);
   y(ii) = p(2);
   z(ii) = p(3);
end
uFOV = uFOV(~isnan(latFOV));
vFOV = vFOV(~isnan(latFOV));
lonFOV = lonFOV(~isnan(latFOV));
latFOV = latFOV(~isnan(latFOV));

fov.u = uFOV;
fov.v = vFOV;
fov.x = x;
```

```
fov.y = y;
fov.z = z;
fov.lat = latFOV;
fov.lon = lonFOV;

PLOT_FIGURE = 0;
if PLOT_FIGURE == 1
   % plot scan volume and unit circle
   figure;
   plot(ESA.uunit,ESA.vunit,ESA.umax,ESA.
   vmax,'linewidth',2);
   axis equal; grid on;
   xlabel('U (sines)'); ylabel('V (sines)');
   title('ESA Scan Volume');
   legend('Visible Space','ESA Scan Volume')

   % plot scan volume and unit circle
   figure; plot(ESA.uunit,ESA.vunit,ESA.umax,ESA.vmax,hor.u,
   hor.v,'linewidth',2);
   hold on;
   hh = fill(fov.u,fov.v,'b');
   alpha(hh,.3)
   grid on; axis equal;
   xlabel('U (sines)'); ylabel('V (sines)');
   title('ESA Scan Volume');
   legend('Visible Space','ESA Scan Volume','Horizon','FOV')
end
return
```

5.6.8   *main_example1.m*

```
%% main example 1
% generates the example orbit in figure 6-2
clear;
close all;

%% define Earth parameters
Earth = defineEarth;

%% define orbit
% inputs to function
altMin = 30000; % min altitude (km)
altMax = 3000; % max altitude (km)
incl = 90; % inlcination (degrees)
omega = 0; % right ascension longitude (degrees)
periapsis = 0; % angle relative to periapsis (degrees)
nRevs = 1; % number of revs to generate
```

```
offsetRevs = 0; % offset in revs relative to t = 0
timeStep = 10; % time increment (seconds)

% define Satellite orbit
Sat = defineOrbit(Earth,altMin,altMax,incl,omega,periapsis,n
Revs,offsetRevs,timeStep);

%% 3D plot of spherical earth with orbit
% generate a spherical earth
[x,y,z] = sphere(200);
x = x*Earth.Re_km;
y = y*Earth.Re_km;
z = z*Earth.Re_km;

figure;
axis equal off
props.FaceColor= 'texture';
props.EdgeColor = 'none';
props.Cdata = Earth.img;
surface(x,y,-z,props);
view(3)
hold on;

% plot orbit
plot3(Sat.x_km_ECI,Sat.y_km_ECI,Sat.z_km_ECI,'k');
plot3(Sat.x_km_ECI(1),Sat.y_km_ECI(1),Sat.z_km_ECI(1),'g*');
```

5.6.9   main_example2.m

```
%% main example 2
% illustrate orbital parameters with 3D plot

clear;
close all;

%% define Earth parameters
Earth = defineEarth;

%% define orbit
% inputs to function
altMin = 10000; % min altitude (km)
altMax = 1000; % max altitude (km)
incl = 45; % inlcination (degrees)
omega = 45; % right ascension longitude (degrees)
periapsis = -50; % angle relative to periapsis (degrees)
nRevs = 1; % number of revs to generate
offsetRevs = 0; % offset in revs relative to t = 0
timeStep = 10; % time increment (seconds)
```

```
% define Satellite orbit
Sat = defineOrbit(Earth,altMin,altMax,incl,omega,periapsis,n
Revs,offsetRevs,timeStep);

%% 3D plot of spherical earth with orbit
% generate a spherical earth
[x,y,z] = sphere(200);
x = x*Earth.Re_km;
y = y*Earth.Re_km;
z = z*Earth.Re_km;

figure;
axis equal off
props.FaceColor= 'texture';
props.EdgeColor = 'none';
props.Cdata = Earth.img;
surface(x,y,-z,props);
view(3)
hold on;

% plot orbit
plot3(Sat.x_km_ECI,Sat.y_km_ECI,Sat.z_km_ECI,'k');
plot3(Sat.x_km_ECI(1),Sat.y_km_ECI(1),Sat.z_km_ECI(1),'g*');
plot3(Sat.x_km_ECI(1),Sat.y_km_ECI(1),Sat.z_km_ECI(1),'go');

% create an equatorial (x-y) plane
planeX = 15*[-1e3 -1e3 1e3 1e3];
planeY = 15*[-1e3 1e3 1e3 -1e3];
planeZ = [0 0 0 0];
hh = fill3(planeX,planeY,planeZ,'c','facealpha',0.1);
fill3(Sat.x_km_ECI,Sat.y_km_ECI,Sat.z_km_
ECI,'m','facealpha',0.2);
line([min(planeX),max(planeX)],[0 0],[0 0]); % x-axis
line([0 0],[min(planeY),max(planeY)],[0 0]); % y-axis
```

5.6.10   main_example3.m

```
%% main example 3
% compute horizon and FOV, and project antenna pattern to ECEF
clear;
close all;

%% define Earth parameters
Earth = defineEarth;

%% define orbit
% inputs to function
altMin = 250; % min altitude (km)
```

```matlab
altMax = 250; % max altitude (km)
incl = 40; % inlcination (degrees)
omega = -100; % right ascension longitude (degrees)
periapsis = 0; % angle relative to periapsis (degrees)
nRevs = 1; % number of revs to generate
offsetRevs = 0; % offset in revs relative to t = 0
timeStep = 10; % time increment (seconds)

% define Satellite orbit
Sat = defineOrbit(Earth,altMin,altMax,incl,omega,periapsis,n
Revs,offsetRevs,timeStep);

%% define ESA parameters
% mechanical orientation of ESA relative to satellite
azMech = 0; % (degrees) aligned with the velocity vector
elMech = 45; % (degrees) pointed 10 degrees up from nadir

% define max grating lobe free scan angles
azMaxScan = 75;
elMaxScan = 50;

ESA = defineESA(azMech,elMech,azMaxScan,elMaxScan);

% % test coordinate systems
% figure;
% quiver3(0,0,0,1,0,0,'color','b'); hold on;
% quiver3(0,0,0,0,1,0,'color','g'); hold on;
% quiver3(0,0,0,0,0,1,'color','r'); hold on;
% xlabel('x'); ylabel('y'); zlabel('z');
% axis equal;
% x = ESA.ant2sat*[0;0;1];
% quiver3(0,0,0,x(1),x(2),x(3),'color','k');

%% time loop
for ii = 1%:length(Sat.time_sec_range)

    % current position
    pos = Sat.pos(:,ii); % current position (ecef)
    llh = Sat.llh(:,ii); % current position (lat/lon/hgt)

    % current coordinate transformations
    xyz2sat = Sat.xyz2sat(:,:,ii);
    sat2xyz = Sat.sat2xyz(:,:,ii);
    ant2sat = ESA.ant2sat;
    sat2ant = ESA.sat2ant;

    % compute horizon
    hor = computeHorizon(Earth,xyz2sat,sat2ant,pos,llh);
```

```
% compute FOV
fov = computeFOV(Earth,ESA,sat2xyz,ant2sat,pos,hor);

% compute boresight point
LOS = sat2xyz*ant2sat*[0;0;1];
[pB,llhB] = los2ecef(pos,Earth.Re_km,LOS);
latBoresight = llhB(1);
lonBoresight = llhB(2);
clear pB llhB
%% plot horizon and FOV
figure;
image(Earth.imgLon,Earth.imgLat,Earth.img);
axis image; axis xy; grid on;
hold on;
fill(fov.lon,fov.lat,'b'); % fill FOV in blue
plot(Sat.lon,Sat.lat,'g.'); % satellite ground track in
green
plot(llh(2),llh(1),'g*'); % current satellite nadir
position
plot(lonBoresight,latBoresight,'w*'); % mechanical
boresight of ESA
plot(hor.lon,hor.lat,'w'); % horizon contour in white
plot(fov.lon,fov.lat,'c'); % FOV contour in cyan

%% interpolate sine space antenna pattern to lat/lon
latpat = linspace(min(hor.lat),max(hor.lat),400);
lonpat = linspace(min(hor.lon),max(hor.lon),400);
[lonpat,latpat] = meshgrid(lonpat,latpat);
[xpat,ypat,zpat] = sph2cart(lonpat*pi/180,latpat*pi/180,E
arth.Re_km);
LOS = cat(3,xpat,ypat,zpat) - permute(repmat(pos,[1,size(
xpat)]),[2,3,1]);
range = sqrt(sum(LOS.^2,3));
LOS = LOS ./ repmat(range,[1,1,3]); % normalize LOS
LOS = reshape(LOS,[size(LOS,1)*size(LOS,2),size(
LOS,3)]).';
LOS = sat2ant * xyz2sat * LOS;
upat = LOS(2,:);
vpat = LOS(1,:);
in1 = inpolygon(upat,vpat,hor.u,hor.v);
iValid = (LOS(3,:) >= 0) & (range(:)' <= hor.range); %
points must be in horizon line, less than
%figure; plot(upat,vpat,'.'); hold on;
plot(hor.u,hor.v,'r'); title('Horizon points');
patInterp = griddata(ESA.u,ESA.v,ESA.pat,upat,vpat);
patInterp(~iValid) = nan;
patInterp = reshape(patInterp,size(latpat));
```

```
%figure; pcolor(lonpat,latpat,patInterp); shading interp;
hold on; plot(hor.lon,hor.lat);

%% plot antenna pattern on map
figure;
image(Earth.imgLon,Earth.imgLat,Earth.img);
axis image; axis xy; grid on;
hold on;
ppp = patInterp - min(patInterp(:));
ppp = ppp/max(ppp(:))*256;
hh = pcolor(lonpat,latpat,ppp); shading interp;
plot(Sat.lon,Sat.lat,'g.'); % satellite ground track in
green
plot(llh(2),llh(1),'g*'); % current satellite nadir
position
plot(lonBoresight,latBoresight,'w*'); % mechanical
boresight of ESA
plot(hor.lon,hor.lat,'w'); % horizon contour in white
plot(fov.lon,fov.lat,'c'); % FOV contour in cyan

end
```

References

Bate, Roger R., Donald D. Mueller, and Jerry E. White. *Fundamentals of Astrodynamics.* New York: Dover Press, 1971.

Vallado, David. *Fundamentals of Astrodynamics and Applications,* 3rd ed. Hawthorne, CA: Microcosm Press, 2007.

chapter six

Electronically Scanned Array Reliability*

Jabberia Miller
Northrop Grumman Electronic Systems

Contents

6.1   Introduction

An electronically scanned array (ESA) provides a significant advantage over its predecessors (passive arrays and reflector antennas) via beam agility. Phase shifters at each radiator element of an ESA provide the capability of rapidly scanning the array beam on the order of microseconds. The ability to accurately command array beams spatially and in time has opened the door to multifunction systems where multiple modes and functions can be interlaced in time or done simultaneously (Skolnik 2008). This enables multiple search volumes to be scanned much faster than a mechanically scanned passive array or reflector antenna could be moved to scan a similar area.

A typical ESA may have an element count on the order of 1,000 elements. This implies 1,000 T/R modules required to operate the ESA. A logical question asked by those unfamiliar with ESAs is: What is the impact of failures on array performance and reliability? This leads to an

---

* This chapter is based upon a technical memo written by Bill Hopwood (2001).

interesting characteristic of ESAs: graceful degradation. An ESA can have failed elements and still operate at a high performance level. This leads to a highly reliable aperture that does not negatively affect the end of life (EOL) of a system. This property of ESAs makes them an invaluable and formidable technology solution where beam agility and multifunction simultaneity are required.

The inherent advantages of an ESA are enabled through architectures that have active components distributed throughout the front end of the system. A transmit and receive (T/R) module can be placed behind every element in the array. The T/R module provides amplification and phase adjustments to the signals that enter and exit the antenna element. The T/R module is controlled by a control module that has drain switches, voltage regulators, digital controls, and memory chips (Agrawal and Holzman 1999). The control module will typically control a group of T/R modules. Failure of multiple T/R modules or control modules will degrade the performance of the array, resulting in higher sidelobes that may exceed the performance requirements. The failures can occur during various points in the operational life of the array, such as storage or while operating in harsh environments. The failures could affect the system availability and ultimately lead to the array being returned for repair. The reliability of a system is its ability to operate while meeting requirements over a specified time. This reliability is often described by the mean time between failures (MTBF) of the T/R and control modules.

This chapter presents a method to calculate the MTBF by using a brute force technique and also using the binomial function. It also shows the effects of failures on the sidelobes of both a one-dimensional linear array and a two-dimensional array. The chapter concludes with an approximation of the effect that T/R module failures have on the radar range equation.

6.2   Probability of Failed Elements

A brute force method can be used to calculate the probability that a certain number of elements in the system will fail. The brute force method assumes that the T/R module is either fully operational or completely nonfunctional. For each element, the probability that the element is working is 0.9, while the probability that the element is not working is 0.1. The method has been used to calculate the probability that four or fewer T/R modules will fail in the simplified four-element system that is shown in Figure 6.1. A truth table showing all of the possible states for the four modules system is shown in Figure 6.2.

The probability that all four modules are working or that there are zero failures is 0.6561, while the probability that all four modules are not working is 0.0001. The cumulative probability for each state is also calculated. The number of calculations required to determine the

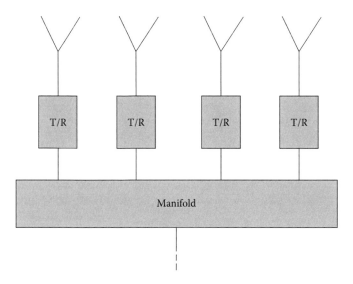

Figure 6.1 Simplified four-element system. Each path through the system consists of an antenna element, T/R module, and path through the manifold. The T/R module could have a phase shifter in addition to amplifiers for transmit and receive. The signals are combined in the manifold. In the case of a system with subarrays, the amplifiers could be placed behind the manifold and used for all four of the elements. Additional manifolds may be used to combine the signals from other subarrays.

cumulative probability for a specific failure state increases as the number of elements is increased, making the brute force method very inefficient for large arrays. The binomial function can be used in these cases. The binomial function, $b(i,N,P)$, calculates the probability of exactly i successes in N trials, when P is the probability of each success. It assumes that the trials are independent. The binomial function is expressed as

$$b(i,N,P) = \frac{N!}{i!(N-i)!} P^i (1-P)^{N-1}$$

(6.1)

The probability of F or fewer successes is then

$$P(\# \text{ of successes} \le F) = \sum_{i=0}^{F} b(i,N,P)$$

(6.2)

which is the sum of the individual probabilities. The binomial function has been used to calculate the probabilities from the previous

	State	Probability	Failures
1	WWWW	.9 .9 .9 .9 = .6561	0
2	WWWF	.9 .9 .9 .1 = .0729	1
3	WW F W	.9 .9 .1 .9 = .0729	1
4	WWFF	.9 .9 .1 .1 = .0081	2
5	WFWW	.9 .1 .9 .9 = .0729	1
6	WFWF	.9 .1 .9 .1 = .0081	2
7	WFF W	.9 .1 .1 .9 = .0081	2
8	WFFF	.9 .1 .1 .1 = .0009	3
9	FWWW	.1 .9 .9 .9 = .0729	1
10	FWWF	.1 .9 .9 .1 = .0081	2
11	FWFW	.1 .9 .1 .9 = .0081	2
12	FWFF	.9 .9 .9 .9 = .6561	3
13	FFWW	.1 .1 .9 .9 = .0081	2
14	FFWF	.1 .1 .9 .1 = .0009	3
15	FFFW	.1 .1 .1 .9 = .0009	3
16	FFFF	.1 .1 .1 .1 = .0001	4

	Number	P_{EACH}	P_{CUM}
States with 0 Failures:	1	0.6561	$1 \times 0.6561 = 0.6561$
States with 1 Failure:	4	0.0729	$4 \times 0.0729 = 0.2916$
States with 2 Failures:	6	0.0081	$6 \times 0.0081 = .0.0486$
States with 3 Failures:	4	0.0009	$4 \times 0.0009 = .0036$
States with 4 Failures:	1	0.0001	$1 \times 0.0001 = 0.0001$

P(4 or Fewer Failures) = .6561 +.29 16+.0486 + .0086 + .000 1 = 1

Figure 6.2 Truth table showing all of the states for the four-element system. A W denotes a working module, while an F denotes a failed module. The cumulative probability for having zero, one, two, three, or four failures is calculated at the bottom of the figure.

four-element system example. The probability that all four of the modules are working is

$$b(4,4,0.9) = \frac{4!\, 0.9^4 (1-0.9)^{4-4}}{4!\,(4-4)!} = 0.9^4 = 0.6561 \tag{6.3}$$

Similarly, the probability that three, two, one, or zero of the four modules are working is

$$b(3,4,0.9) = \frac{4!\, 0.9^3 (1-0.9)^{4-3}}{3!\,(4-3)!} = \frac{4*3*2*1}{3*2*1} 0.9^3\, 0.1 = 0.2916 \tag{6.4}$$

$$b\,(2,4,0.9) = \frac{4!\,0.9^2\,(1-0.9)^{4-2}}{2!\,(4-2)!} = \frac{4*3*2*1}{2*1*2*1}0.9^2\,0.1^2 = 0.0486 \qquad (6.5)$$

$$b\,(1,4,0.9) = \frac{4!\,0.9^1\,(1-0.9)^{4-1}}{1!\,(4-1)!} = \frac{4*3*2*1}{1*3*2*1}0.9\,0.1^3 = 0.0036 \qquad (6.6)$$

$$b\,(0,4,0.9) = \frac{4!\,0.9^0\,(1-0.9)^{4-0}}{0!\,(4-0)!} = \frac{4*3*2*1}{1*4*3*2*1}1\,0.1^4 = 0.0001 \qquad (6.7)$$

$$0.6561 + 0.2916 + 0.0486 + 0.0036 + 0.0001 = 1 \qquad (6.8)$$

The sum of the five probabilities is equal to 1, as previously shown in Figure 6.2. The binomial function is an efficient means of calculating the probability of the number of working modules for a large antenna that may have thousands of elements.

6.3   Mean Time between Failure (MTBF)

The MTBF can be represented as

$$MTBF = \frac{\text{Total Operating Time with Large Population}}{\text{Total Failures During that Time}} \qquad (6.9)$$

The MTBF is the average time between failures of a system. The MTBF for a T/R module can be improved by increasing the integration of MMICs, improving the manufacturing processes, or reducing the module operational temperature (Agrawal and Holzman 1999). If 200 units have been operating for 10,000 hours with 5 total failures, the MTBF can be calculated as

$$MTBF = \frac{200*10,000}{5} = 400,000\,Hours \qquad (6.10)$$

The MTBF can also be calculated from the reliability function, $R(t)$, as shown in Equation (6.11).

$$MTBF = \int_0^{\infty} R(t)\,dt \qquad (6.11)$$

where $R(t)$ can be defined as

$$R(t) = e^{-\lambda t} \qquad (6.12)$$

This leads to

$$MTBF = \int_0^\infty e^{-\lambda t}\, dt = \frac{1}{\lambda}$$ (6.13)

We now find the array MTBF, which is arbitrarily defined by the presence of F or fewer elements in N using

$$R_A(t) = b(i, N, R_E(t))$$ (6.14)

or

$$R_A(t) = \sum_{i=0}^{F} b(i, N, R_E(t))$$ (6.15)

where $R_A(t)$ is the array reliability function and $R_E(t)$ is the element (or module) reliability function $= e^{-\lambda t}$. The array MTBF is

$$MTBF_A = \int_0^\infty R_A(t)\, dt$$ (6.16)

Recall that we calculate the probability of failed elements, not working elements. Thus, we interchange P and $1-p$ in the binomial function. The MTBF becomes

$$MTBF_A = \sum_{i=0}^{F} \int_0^\infty \frac{N!}{i!(N-i)!} e^{-\lambda t(N-i)} (1 - e^{-\lambda t})\, dt$$ (6.17)

Upon manipulation, this reduces to

$$MTBF_A \equiv \frac{1}{\lambda_E} \sum_{i=0}^{F} \frac{1}{N-i} \approx \frac{1}{\lambda_E} \frac{F}{N} = MTBF_E \frac{F}{N}$$ (6.18)

Thus, the array MTBF scales the element MTBF by the fraction that can fail. For example, if the element MTBF is 1e6 hours and λ_E is 1e-6, and if we tolerate 5% failures ($F/N = 0.05$), the array MTBF is 0.05*1e6 = 50,000 hours. The array MTBF can be described as

$$MTBF_A = \frac{1}{\lambda_E} \frac{F}{N}$$ (6.19)

where λ_E is the element failure rate and $\frac{F}{N}$ is the allowable failure quantity.

6.4 Effects of Module Failures on 1D and 2D Antenna Patterns

One of the advantages of distributing the modules throughout the system is a graceful degradation of the performance over time. The probability of all the modules failing at the same time is low. As the failures occur, the sidelobe levels rise while the gain generally drops. The ability to model the effects of the failures is useful during the design and testing phases of the system. This section presents MATLAB® code that calculates antenna patterns for cases where a random number of elements fail or when the system has subarrays and a subarray amplifier fails.

For this system, the analysis is done at 3.85 GHz. The system has 1,600 elements that are spaced $l/2$ apart on a rectangular grid. The locations of the elements are show in Figure 6.3. The antenna element pattern is modeled as $\cos^{1.35}\theta$. This pattern is shown in Figure 6.4. The normalized two-dimensional (2D) antenna pattern is shown in Figure 6.5 along with azimuth and elevation scans that are shown in Figures 6.6 and 6.7. The patterns show that the peak of the nearest sidelobe to the main beam is 13 dB lower, as theory suggests.

An example is shown in Figures 6.8 to 6.11 where 10% of the elements in the system are randomly allowed to fail. The elements are shown in Figure 6.8. The failed elements are circled. The resulting 2D

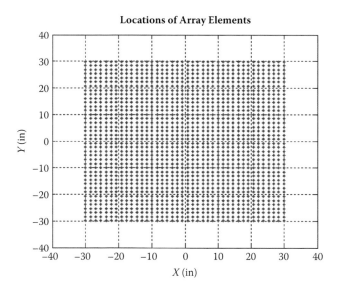

Figure 6.3 40 × 40 grid of elements used in the analysis. The elements are spaced $l/2$ apart.

Element Pattern, 40 × 40 Elements

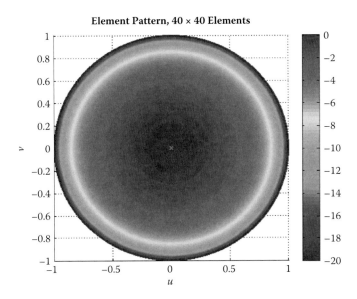

Figure 6.4 Pattern for element that has cos1.35Θ roll-off. The x represents the scan angle. In this case, the antenna's beam has not been scanned. The element pattern does not move when the antenna pattern is scanned.

Pattern, 40 × 40 Elements

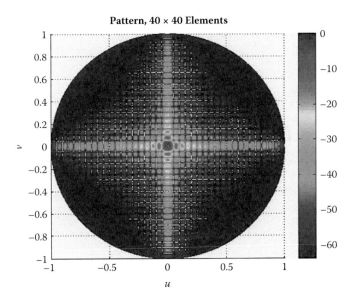

Figure 6.5 Antenna pattern for 40 × 40 grid of elements spaced *l*/2 apart. The pattern has been normalized to the peak of the beam. See insert.

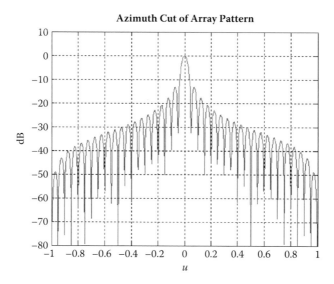

Figure 6.6 One-dimensional azimuth cut for 2D pattern shown in Figure 6.5. Notice that the sidelobes are 13 dB lower than the peak of the beam.

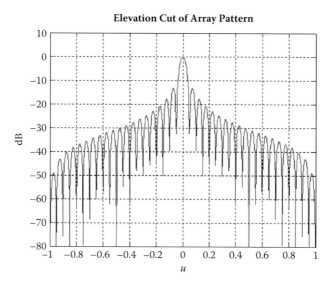

Figure 6.7 One-dimensional elevation cut for 2D pattern shown in Figure 6.5. Notice that the sidelobes are 13 dB lower than the peak of the beam.

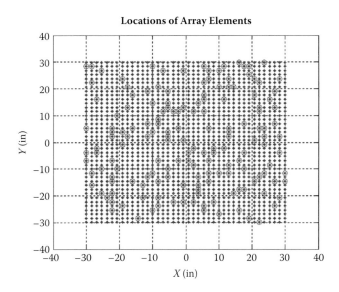

Figure 6.8 Ten percent of the elements have failed. The failed elements are circled in red. See insert.

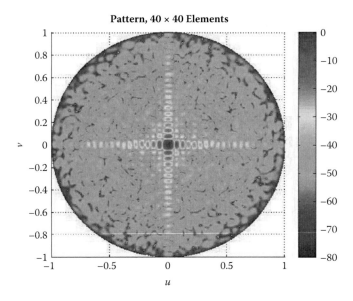

Figure 6.9 Two-dimensional pattern of antenna with 10% failures. See insert.

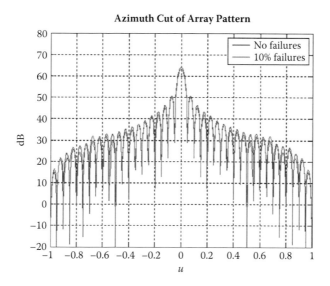

Figure 6.10 Comparison of one-dimensional (1D) azimuth cuts for antenna without failures and one with 10% failures. The patterns have not been normalized to show the reduction in gain as elements are allowed to fail. An elevation cut would show a similar result.

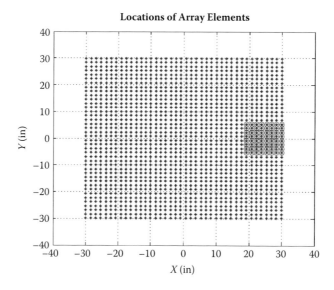

Figure 6.11 Element failures in a subarray. The circled elements are not working in this analysis.

pattern is shown in Figure 6.9. The intercardinal areas of the pattern are higher than the baseline 2D case without any failures, as shown in Figure 6.5. The azimuth cut plot of Figure 6.10 has not been normalized. Examining the figure at $u = 0$ shows the reduction in gain due to the failures.

Next, performance in the presence of failures is analyzed for the case where the example ESA architecture is divided into subarrays where the subarrays are comprised of eight elements in both azimuth and elevation. This leads to a total of five subarrays in azimuth and five subarrays in elevation. Moving the amplifiers behind the subarrays reduces the system costs as fewer components are required. If a subarray amplifier fails, all of the elements in the subarray may not be able to function. The patterns for this case are shown below. The elements of the system are shown in Figure 6.11. The elements that are circled represent the failed elements. The resulting 2D pattern is shown in Figure 6.12. An azimuth principal plane cut does not deviate much from the baseline case without failures. The majority of the changes in the pattern occur in the intercardinal planes. Additional patterns can be produced by making modifications to the MATLAB script that is in Section 6.6.1.

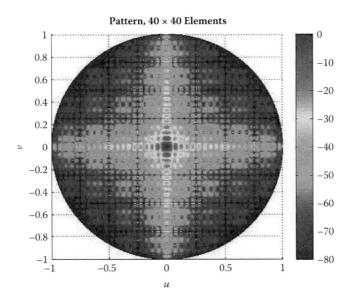

Figure 6.12 Two-dimensional pattern from the array with a failed subarray as described in Figure 11. See insert.

6.5 Effect of Module Failures on the Radar Range Equation

The radar range equation is used to estimate the range of a radar as a function of the radar characteristics (Skolnik 2008). The equation can be used during design and system analysis stages. The radar range equation can also be studied to show the relationship that failed modules have on the radar detection range. Consider a system that has F failed elements out of N total elements. The number of live elements is then

$$Live\ Elements = N - F \tag{6.20}$$

The failures destroy the signal for both transmit and receive. The transmit power becomes

$$Tx\ Power = P_E(N - F) \tag{6.21}$$

The transmit gain is scaled by the number of live elements and is represented as

$$Tx\ Gain = \frac{4\pi A}{\lambda^2} = \frac{4\pi}{\lambda^2} A_E(N - F) \tag{6.22}$$

The transmit effective radiated power (ERP) is the transmit gain multiplied by the transmit radiated power and is written as

$$Tx\ ERP = P_E(N - F)\frac{4\pi}{\lambda^2} A_E(N - F) \tag{6.23}$$

The power density at range R is expressed as

$$Power\ Density\ at\ Range\ R = P_E\frac{4\pi A_E}{\lambda^2}\frac{(N - F)^2}{4\pi R^2} \tag{6.24}$$

The antenna transmits a signal that hits a target and returns to the antenna. The power density at the face of the receiver is

$$Power\ Density\ at\ RCVR = P_E\frac{4\pi A_E}{\lambda^2}\frac{(N - F)^2}{4\pi R^2}\frac{\sigma}{4\pi R^2}\frac{w}{m^2} \tag{6.25}$$

The reflected signal that arrives at the face of the array enters the array through the antenna element that has a defined area and travels through the LNA and manifolds before being processed in the receiver. The power received in an element of area A_E, after the LNA gain (G_{LNA}) can be represented as

$$Power\ Received\ in\ Area\ A_E = \frac{P_E\,A_E^2\,(N-F)^2\,\sigma\,G_{LNA}}{4\pi R^4\lambda^2} \tag{6.26}$$

The manifold immediately follows the LNA, and the RMS manifold voltage without failures is

$$RMS\ Manifold\ Voltage = V_M = \sqrt{\frac{Z_o\,P_E\,A_E^2\,(N-F)^2\,\sigma\,G_{LNA}}{4\pi R^4\lambda^2}} \tag{6.27}$$

The voltage at the manifold output due to a single live element is expressed as

$$V_{M_e} = \frac{1}{\sqrt{N}}\sqrt{\frac{Z_o\,P_E\,A_E^2\,(N-F)^2\,\sigma\,G_{LNA}}{4\pi R^4\lambda^2}} \tag{6.28}$$

The coherent (tuned) voltage sum due to $(N-F)$ live elements at the manifold output is denoted as

$$V_{M_{out}} = \frac{N-F}{\sqrt{N}}\sqrt{\frac{Z_o\,P_E\,A_E^2\,(N-F)^2\,\sigma\,G_{LNA}}{4\pi R^4\lambda^2}} \tag{6.29}$$

The manifold output signal power is

$$P_{M_{out}} = \frac{(N-F)^2}{N}\frac{P_E\,A_E^2\,(N-F)^2\,\sigma\,G_{LNA}}{4\pi R^4\lambda^2} \tag{6.30}$$

This expression can be simplified to

$$P_{M_{out}} = \frac{(N-F)^4}{N}\frac{P_E\,A_E^2\,\sigma\,G_{LNA}}{4\pi R^4\lambda^2}\quad(\alpha N^3) \tag{6.31}$$

where the result is proportional to N^3. The manifold output average power becomes

$$\left(P_{M_{out}}\right)_{AVG} = \frac{(N-F)^4}{N}\frac{P_E\,A_E^2\,\sigma\,G_{LNA}}{4\pi R^4\lambda^2}\frac{T_P}{T_{PP}} \tag{6.32}$$

where $\frac{T_P}{T_{PP}}$ represents the duty cycle.

The noise power density in the radar range equation has three components: the sky temperature (T_{SKY}), LNA temperature (T_{LNA}), and LNA gain (G_{LNA}). The noise at the LNA output is

$$Noise\ at\ LNA\ Output = k(T_{SKY} + T_{LNA})G_{LNA}\ \ Watts/Hz \qquad (6.33)$$

and is equal to the noise at the manifold input. Manifold output noise due to one live element is

$$Noise\ at\ Manifold\ Output_{1\ element} = k(T_{SKY} + T_{LNA})\frac{G_{LNA}}{N} \qquad (6.34)$$

The manifold output noise due to $(N - F)$ elements is represented as

$$Noise\ at\ Manifold\ Output_{N-F\ elements} = k(T_{SKY} + T_{LNA})G_{LNA}\ \frac{N-F}{N} \qquad (6.35)$$

We make the assumption that receiver noise after manifold is second order and does not contribute. The signal-to-noise ratio (SNR) is defined as

$$SNR = \frac{Average\ Signal\ Power}{Noise\ in\ PRF\ Bandwidth\ (1/T_{PP})} \qquad (6.36)$$

where $T_{PP} = (1/PRF)$. Substituting Equations (6.32) and (6.35) into Equation (6.36) yields

$$SNR = \frac{\frac{(N-F)^4}{N}\ \frac{P_E\ A_E^2\ \sigma G_{LNA}}{4\pi R^4}\ \frac{T_P}{T_{PP}}}{k(T_{SKY} + T_{LNA})G_{LNA}\ \frac{N-F}{N}\ \frac{1}{T_{PP}}} \qquad (6.37)$$

The SNR in Equation (6.37) can be reduced to

$$SNR = \frac{(N-F)^3}{N^3}\ \frac{P_E\ A_E^2\ \sigma T_P\ N^3}{k(T_{SKY} + T_{LNA})4\pi R^4} \qquad (6.38)$$

For a fixed SNR, R^4 is denoted by

$$R^4 = \left(1 - \frac{F}{N}\right)^3\ \frac{P_E\ A_E^2\ \sigma T_P\ N^3}{k(T_{SKY} + T_{LNA})4\pi} \qquad (6.39)$$

The ratio of R^4 with failures to R^4 without failures is equal to

$$\frac{R^4 \text{ with Failures}}{R^4 \text{ without Failure}} = \left(1 - \frac{F}{N}\right)^3 \tag{6.40}$$

Using the following substitution,

$$\frac{F}{N} = \lambda_e t \qquad (t \text{ in hours}) \tag{6.41}$$

Equation (6.40) becomes

$$\frac{R \text{ with Failures}}{R \text{ without Failure}} = (1 - \lambda_E t)^{3/4} \tag{6.42}$$

where $t = MTBF_A$. Equation (6.42) shows that the radar's range would be severely degraded when $t = 1/\lambda_E$, which is when the modules are largely dying.

Different modes of failures exist for the T/R module. The transmit power, transmit gain, and receive gain could all fail. This would lead to the range performance decreasing by $(1 - F/N)^{3/4}$. If the module's transmit functionality (both transmit power and gain) fails while the receive portion of the module is working, the range performance drops as $(1 - F/N)^{2/4}$.

6.6 MATLAB Program Listings

This section contains a listing of all MATLAB programs and functions used in this chapter.

6.6.1 Reliability Code

```
%Pattern code for an mXn Array
%Element and Subarray Failures can be evaluated
%Written by: Jabberia Miller

clear all;clc
%% Variables
f=3.85e9; % Frequency (GHz)
c=11.80285e9; %Speed of Light (in/s)
lambda=c/f;
k=2*pi/lambda; %wavenumber
omega=2*pi*f;

elemfac=1.35; %Element Factor
theta_scan = 0; %Theta Scan Angle (degrees)
phi_scan = 0; %Phi Scan Angle (degrees)
```

```
%Number of Elements in x and y
M=40; %number of x elements
N=40; %number of y elements

subM = 5; %number of subarrays in x - must be exactly divis-
ible by num of elements
subN = 5; %number of subarrays in y - must be exactly
divisible by num of elements

removeElem = 2; % 0 - no failures
          % 1 - fail random elements
          % 2 - fail random subarray
percentFail=10; % percent of elements to fail

%%

%Compute Spacing between elements
dx=lambda/2;
dy=lambda/2;

%Compute Max Gain
A=(M*dy)*(N*dx);
G=4*pi*A/(lambda^2);

%Grid of locations for element x and y
xtemp = ([1:M] - 0.5*(M+1))*dx;
ytemp = ([1:N] - 0.5*(N+1))*dy;
[x y] = meshgrid(xtemp,ytemp);

%Grid of locations for subarray x and y
subdx = M/subM ; %number of elements per subarray in x
subdy = N/subN ; %number of elements per subarray in y
subtempx = ([1:subM] - 0.5*(subM+1))*subdx*dx;
subtempy = ([1:subN] - 0.5*(subN+1))*subdy*dy;
[subx suby] = meshgrid(subtempx,subtempy);

%Associate Elements with subarrays
subnum = subM*subN; %Calculate number of Subarrays
sub = ones(size(reshape(x,M*N,1))); %storage variable
x1=reshape(x,M*N,1); % reshape x array into vector
y1=reshape(y,M*N,1); % reshape y array into vector
offst=0.1;
for ict = 1:subnum

  jj = find(x1>=(subx(ict)-subdx*dx/2-offst) & x1<=(subx(ict)
  +subdx*dx/2+offst) & y1 <=(suby(ict)+subdy*dy/2+offst) & y1
  >=(suby(ict)-subdy*dy/2-offst));
  sub(jj) = ict;
```

```
end
sub = reshape(sub,M,N);

%compute angles where pattern will be evaluated over
[Az,El] = meshgrid(-90:0.2:90,-90:0.2:90);

%convert to Az/EL angles to theta/phi
Phyr = Az * pi/180; %convert to radians
Thyr = (90 - El) * pi/180; %convert to radians

u = sin(Thyr) .* cos(Phyr);
v = sin(Thyr) .* sin(Phyr);
w = cos(Thyr);
thetamat = acos(u ./ sqrt(v.^2 + w.^2 + u.^2)) ;
phimat = atan2(w,v) ;

%Compute direction cosines
u=sin(thetamat).*cos(phimat);
v=sin(thetamat).*sin(phimat);

%Compute steering angles
thetao=theta_scan*pi/180;
phio=-phi_scan*pi/180;
uo=sin(thetao)*cos(phio);
vo=sin(thetao)*sin(phio);

%Remove elements or subarrays
AF=zeros(size(u));
elempat=zeros(size(u));
wgts = ones(size(x));
switch removeElem
  case 1 %remove random elements
    numdelete = round(percentFail*M*N/100);
    DeleteList = randperm(M*N);
    DeleteList = DeleteList(1:numdelete);
    wgts(DeleteList)= 0;
  case 2 % remove subarray
    deletesub = randperm(subM*subN);
    [tmpx tmpy] = find(sub == deletesub(1));
    wgts(tmpx,tmpy)=0;
end

%Compute Array Factor
for ii=1:M*N
    tm = (x(ii)*u + y(ii)*v)/c;
    tmo = (x(ii)*uo + y(ii)*vo)/c;
    AF = AF + wgts(ii)*exp(j*omega*(tm-tmo));
end
```

```
%Compute element pattern
elempat=cos(thetamat).^elemfac;
elempatdB=20*log10(abs(elempat+eps));

%Compute Pattern
Pattern=AF.*elempat;

Patternmag=abs(Pattern);
Patternmagnorm=Patternmag/max(max(Patternmag));
%normalized pattern
PatterndB=20*log10(Patternmag+eps);
PatterndBnorm=20*log10(Patternmagnorm+eps);

%Set floor
dBfloor=-80;
PatterndB(find(PatterndB < dBfloor))=dBfloor;
PatterndBnorm(find(PatterndBnorm < dBfloor))=dBfloor;

%Generate Figures
figure 1)
clf
surf(u,v,PatterndB);hold on
plot(uo,vo,'yx');
caxis([-50 20*log10(M*N)+5])
view(0,90),shading interp,colorbar
title(['Pattern,',num2str(M),'x',num2str(N),' Elements']);
xlabel('u')
ylabel('v')

figure 11)
clf
surf(u,v,PatterndBnorm);hold on
plot(uo,vo,'yx');
caxis([-80 0])
view(0,90),shading interp,colorbar
title(['Pattern,',num2str(M),'x',num2str(N),' Elements']);
xlabel('u')
ylabel('v')

[row col]=find(PatterndB==max(max(PatterndB))); %find row
and column of max
figure 2);clf;
plot(u(row,:),PatterndB(row,:),'b-');
grid on;
title('Azimuth Cut of Array Pattern');
xlabel('u');
ylabel('dB');
axis([-1 1 -20 80]);
```

```
figure 3);clf;
plot(v(:,col),PatterndB(:,col),'b-');
grid on;
title('Elevation Cut of Array Pattern');
xlabel('u');
ylabel('dB');
axis([-1 1 -20 80]);

figure 4)
clf
surf(u,v,elempatdB),hold
caxis([-20 0])
view(3),shading interp,colorbar
title(['Element Pattern,',num2str(M),'x',num2str(N),'
Elements'])
plot(uo,vo,'yx')
view(0,90);
xlabel('u')
ylabel('v')

figure 5)
clf
plot(x,y,'b.');grid on;hold on;
switch removeElem
  case 1
    plot(x(DeleteList),y(DeleteList),'ro')
  case 2
    [xplot yplot]=find(wgts==0);
    plot(x(xplot,yplot),y(xplot, yplot),'ro')
end
title('Locations of Array Elements')
xlabel('X (in) ');
ylabel(' Y (in) ');
axis([-40 40 -40 40]);

return
```

References

Agrawal, A. K., and E. L. Holzman. Active Phased Array Design for High Reliability. *IEEE Transactions—Aerospace and Electronic Systems*, 35, no. 4 (1999): 1204–1210.

Hopwood, F. W. *Effects of Module Reliability on ESA Reliability and Radar Performance.* Technical Memo. Baltimore, MD: Northrop Grumman Electronic Systems, August 2001.

Skolnik, Merrill. *Radar Handbook.* McGraw-Hill, New York, NY, 2008.

Appendix A: Array Factor (AF) Derivation

In Chapter 1 the closed form for the AF was shown to be

$$AF = \frac{\sin\left[M\pi d\left(\frac{\sin\theta_o}{\lambda_o} - \frac{\sin\theta}{\lambda}\right)\right]}{\sin\left[\pi d\left(\frac{\sin\theta_o}{\lambda_o} - \frac{\sin\theta}{\lambda}\right)\right]}$$

(A.1)

This is derivable from the exponential summation expression for the uniform illumination AF, which is

$$AF = \sum_{m=1}^{M} e^{j\left(\frac{2\pi}{\lambda}x_m \sin\theta - \frac{2\pi}{\lambda_o}x_m \sin\theta_o\right)}$$

(A.2)

The position of the array elements, x_m, is expressed as $x_m = \left(m - \frac{M+1}{2}\right)d_x$, where d_x is the element spacing and M is the number of array elements. Using this expression puts the phase center of the array at $x = 0$.[*] Equation (A.2) can then be rewritten as

$$AF = \sum_{m=1}^{M} e^{j\left(\frac{2\pi}{\lambda}\sin\theta - \frac{2\pi}{\lambda_o}\sin\theta_o\right)x_m}$$

$$= \sum_{m=1}^{M} e^{jd_x\left(\frac{2\pi}{\lambda}\sin\theta - \frac{2\pi}{\lambda_o}\sin\theta_o\right)\cdot\left(m - \frac{M+1}{2}\right)}$$

(A.3)

$$= \sum_{m=1}^{M} e^{j\Psi\left(m - \frac{M+1}{2}\right)}$$

[*] The expression for x_m can be represented as $x_m = (m - 1)d_x$ without any loss of generality; however, this gives a phase center that is offset from $x = 0$.

where $\Psi = d_x(\frac{2\pi}{\lambda}\sin\theta - \frac{2\pi}{\lambda_o}\sin\theta_o)$. Equation (A.3) can then be expanded as

$$AF = \sum_{m=1}^{M} e^{j\Psi\left(m-\frac{M+1}{2}\right)}$$

(A.4)

$$= e^{j\Psi\left(\frac{1-M}{2}\right)} + e^{j\Psi\left(\frac{3-M}{2}\right)} + \cdots + e^{j\Psi\left(\frac{M-3}{2}\right)} + e^{j\Psi\left(\frac{M-1}{2}\right)}$$

and when scaled by $e^{j\Psi}$ results in

$$e^{j\Psi}AF = e^{j\Psi\left(\frac{3-M}{2}\right)} + \cdots + e^{j\Psi\left(\frac{M-1}{2}\right)} + e^{j\Psi\left(\frac{M+1}{2}\right)}$$

(A.5)

Subtracting Equation (A.5) from Equation (A.4) generates the following expression

$$AF - e^{j\Psi}AF = AF(1-e^{j\Psi}) = e^{j\Psi\left(\frac{1-M}{2}\right)} - e^{j\Psi\left(\frac{M+1}{2}\right)}$$

(A.6)

Rearranging terms in Equation (A.6) results in a new expression for the *AF*, which is

$$AF = \frac{e^{j\Psi\left(\frac{1-M}{2}\right)} - e^{j\Psi\left(\frac{M+1}{2}\right)}}{(1-e^{j\Psi})}$$

(A.7)

Equation (A.7) can then be further simplified as follows:

$$AF = \frac{e^{j\Psi\left(\frac{1-M}{2}\right)} - e^{j\Psi\left(\frac{M+1}{2}\right)}}{(1-e^{j\Psi})}$$

$$= \frac{e^{j\left(\frac{\Psi}{2}-\frac{\Psi M}{2}\right)} - e^{j\left(\frac{\Psi}{2}+\frac{\Psi M}{2}\right)}}{e^{j\frac{\Psi}{2}}\left(e^{-j\frac{\Psi}{2}} - e^{j\frac{\Psi}{2}}\right)}$$

(A.8)

$$= \frac{e^{j\frac{\Psi}{2}}\left(e^{-j\Psi\left(\frac{M}{2}\right)} - e^{j\Psi\left(\frac{M}{2}\right)}\right)}{e^{j\frac{\Psi}{2}}\left(e^{-j\frac{\Psi}{2}} - e^{j\frac{\Psi}{2}}\right)}$$

$$= \frac{e^{-j\Psi\left(\frac{M}{2}\right)} - e^{j\Psi\left(\frac{M}{2}\right)}}{e^{-j\frac{\Psi}{2}} - e^{j\frac{\Psi}{2}}}$$

Using Euler's identity, $\sin\theta = \frac{e^{j\theta}-e^{-j\theta}}{2j}$, Equation (A.8) can then be reduced to the following expression for the *AF*:

$$AF = \frac{\sin\left(M \cdot \frac{\Psi}{2}\right)}{\sin\left(\frac{\Psi}{2}\right)} \tag{A.9}$$

Substituting for Ψ we're left with

$$AF = \frac{\sin\left[M \cdot d_x\left(\frac{\pi}{\lambda}\sin\theta - \frac{\pi}{\lambda_o}\sin\theta_o\right)\right]}{\sin\left[d_x\left(\frac{\pi}{\lambda}\sin\theta - \frac{\pi}{\lambda_o}\sin\theta_o\right)\right]} \tag{A.10}$$

which is equivalent to Equation (A.1).

Appendix B: Instantaneous Bandwidth (IBW) Derivation

In Chapter 1, the expression for IBW was shown to be

$$IBW = \frac{kc}{L\sin\theta_o} \tag{B.1}$$

where k is the beamwidth factor, L is the length of the ESA, and θ_o is the maximum required scan angle. An alternative expression can be shown that ultimately reduces to Equation (B.1). We begin with the expression for the AF shown in Chapter 1:

$$AF = \sum_{m=1}^{M} a_m e^{j\left(\frac{2\pi}{\lambda}x_m\sin\theta - \frac{2\pi}{\lambda_o}x_m\sin\theta_o\right)} \tag{B.2}$$

Rewriting Equation (B.2) in terms of f, the AF expression becomes

$$AF = \sum_{m=1}^{M} a_m e^{j\frac{2\pi}{c}x_m(f\sin\theta - f_o\sin\theta_o)} = \sum_{m=1}^{M} a_m e^{j\frac{2\pi}{c}x_m(\Psi)} \tag{B.3}$$

When operating at the tune frequency, $f = f_o$, and the ESA is scanned to $\theta = \theta_o$, $\Psi = 0$. However, for operation away from the tune frequency, $f = f_o + \Delta f$, $\Psi = (f_o + \Delta f)\sin\theta - f_o\sin\theta_o$ and Ψ is no longer zero, thus resulting in beam squint.

The beam squint, $\Delta\theta$, can be calculated by substituting $f = f_o + \Delta f$ and $\theta = \theta_o + \Delta\theta$ into the expression for Ψ

$$\Psi = f\sin\theta - f_o\sin\theta_o$$
$$= (f_o + \Delta f)\sin(\theta_o - \Delta\theta) - f_o\sin\theta_o \tag{B.4}$$

It is important to note that for $\Delta f > 0$, $\Delta\theta < 0$, meaning the beam squints to a value less than the scan angle, for off-tune frequencies greater than the tune frequency, and a value greater than the scan angle, for off-tune frequencies less than the tune frequency. Setting Ψ equal to 0 in Equation (B.4), applying a trigonometric identity ($\sin(A - B) = \sin A \cos B - \sin B \cos A$) and solving for $\Delta\theta$ produces the following expression:

$$(f_o + \Delta f)[\sin\theta_o \cos\Delta\theta - \sin\Delta\theta \cos\theta_o] = f_o \sin\theta_o \qquad (B.5)$$

Using the small angle approximation ($\sin\alpha \approx \alpha$) for the beam squint, $\Delta\theta$, reduces Equation (B.5) to

$$\Delta\theta = \frac{\Delta f}{(f_o + \Delta f)}\tan\theta_o$$

$$\approx \frac{\Delta f}{f_o}\tan\theta_o \qquad (B.6)$$

The expression in Equation (B.6) is the same as what is found in Skolnik (1990). Recognizing that Δf is the *IBW*, and substituting the expression for the beamwidth for the beam squint, we arrive at the following equation:

$$IBW = \frac{k\lambda}{L} \cdot \frac{f_o}{\tan\theta_o}$$

$$= \frac{kc}{L\tan\theta_o} \qquad (B.7)$$

For scan angles less than 20° Equation (B.7) is identical in value to Equation (B.1).

Appendix C: Triangular Grating Lobes Derivation

In Chapter 2, the grating lobe expression for a triangular grid of elements was shown to be (Skolnik 1990; Corey 1985):

$$u_m = u_o + m\frac{\lambda}{2d_x}, \quad v_n = v_o + n\frac{\lambda}{2d_y}$$

$$m, n = 0, \pm 1, \pm 2,\ldots$$

$$m + n \text{ is even}$$

(C.1)

This expression is derived in Corey (1985); however, a more intuitive derivation will be shown here. We begin by considering a rectangular array with element spacing $2d_x$ and $2d_y$, as shown in Figure C.1. In Chapter 2 it was shown that the grating lobes for this array occur at

$$u_m = u_o + m\frac{\lambda}{(2d_x)}, \quad m = 0, \pm 1, \pm 2,\ldots$$

$$v_n = v_o + n\frac{\lambda}{(2d_y)}, \quad n = 0, \pm 1, \pm 2,\ldots$$

(C.2)

Next we consider another array of elements with the same $2d_x$ and $2d_y$ spacing, but offset by d_x and d_y, as shown in Figure C.2. This offset can be expressed mathematically with a complex exponential phase shift $e^{-j\frac{2\pi}{\lambda}[d_x(u-u_o)+d_y(v-v_o)]}$. The combined array factor (AF_{total}) for the two offset rectangular arrays can be expressed as

$$AF_{total} = AF_1 + AF_2$$

$$= AF_1 + e^{-j\frac{2\pi}{\lambda}[d_x(u-u_o)+d_y(v-v_o)]}AF_1$$

$$= \left(1 + e^{-j\frac{2\pi}{\lambda}[d_x(u-u_o)+d_y(v-v_o)]}\right)AF_1$$

(C.3)

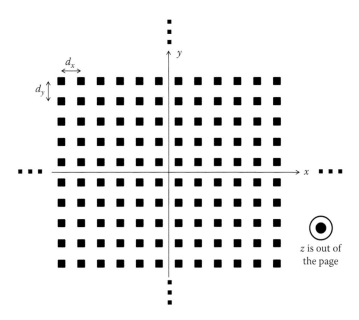

Figure C.1 Rectangular array with element spacing $2d_x$ and $2d_y$.

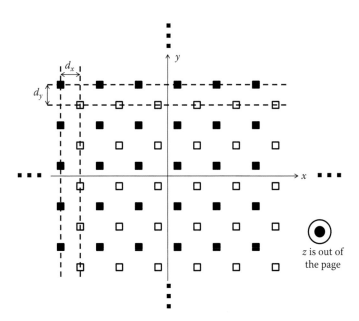

Figure C.2 Rectangular array offset from the array in Figure A.3.1 by d_x and d_y.

In Equation (C.3), AF_1 and AF_2 are the array factors for the two arrays respectively. From Equation (C.2), the AF in Equation (C.3) has maxima (i.e., grating lobes) that occur at multiples of $m\frac{\lambda}{2d_x}$ and $n\frac{\lambda}{2d_y}$. Substituting into the complex exponential phase shift term in Equation (C.3) gives

$$\left(1 + e^{-j\frac{2\pi}{\lambda}\left[d_x\left(m\frac{\lambda}{2d_x}\right) + d_y\left(n\frac{\lambda}{2d_y}\right)\right]}\right) = 1 + e^{-j\pi[m+n]}$$

$$= 0, \text{ for } [m+n] \text{ odd}$$

$$= 1, \text{ for } [m+n] \text{ even}$$

(C.4)

From Equation (C.4) we see that the AF expression in Equation (C.3) has maxima only when the quantity $[m + n]$ is even. These maxima are the grating lobes for a triangular grid that is a superposition of two rectangular grids spatially offset. The expression for the grating lobe locations in sine space is then

$$u_m = u_o + m\frac{\lambda}{2d_x}, \qquad m = 0, \pm1, \pm2,\ldots$$

$$v_n = v_o + n\frac{\lambda}{2d_y}, \qquad n = 0, \pm1, \pm2,\ldots$$

(C.5)

for $[m+n]$ even

Equation (C.5) is equivalent to Equation (C.1).

Index

Printed and bound by CPI Group (UK) Ltd, Croydon, CR0 4YY

18/10/2024

01776271-0005